Foundation

Of

Physics

A Comprehensive Journey

Through the Laws of the Universe

By

Matthew Meyer

Published by Matthew Meyer

Cover design by Matthew Meyer

Edited by Matthew Meyer

Contents

Preface

Physics is the story of the universe — the grand narrative written in the language of mathematics, motion, and energy. It is humanity's effort to understand the invisible threads that bind together the stars in the heavens, the atoms in our hands, and the thoughts in our minds. It reveals not only how the universe works, but also why it inspires such wonder within us.

From the earliest philosophers who gazed at the sky and asked *why things fall*, to the scientists who built equations to describe the dance of galaxies and particles alike, physics has always been our bridge between curiosity and understanding. It transforms mystery into meaning, showing that beneath the complexity of existence lies a profound and elegant order.

This book, *Foundations of Physics: A Comprehensive Journey Through the Laws of the Universe*, was written for students, dreamers, and thinkers who wish to see the world through the lens of natural law. It begins with the timeless principles of classical mechanics — the motion of bodies and the forces that govern them — and guides the reader through the evolution of physics into the realms of thermodynamics, electromagnetism, relativity, quantum mechanics, and the cosmic frontier of astrophysics. Each concept builds upon the last, weaving a continuous thread from falling apples to expanding galaxies, from the heat of stars to the quanta of light.

My goal is to present physics not as a set of formulas to memorize, but as a living, breathing pursuit of truth — a dialogue between observation and imagination. The equations here are not merely symbols; they are the music of the cosmos, the code that underlies all matter and motion. In mastering them, we come to see that science is not opposed to wonder, but the deepest expression of it.

In writing this book, I hope to make physics accessible, engaging, and deeply human — to show that every principle, from Newton's laws to quantum uncertainty, is part of a single, majestic tapestry. Whether you are encountering these ideas for the first time or returning to them with renewed curiosity, may this journey remind you that knowledge is the most beautiful adventure of all.

Let us explore the universe together — not as distant observers, but as participants in its unfolding story. For in understanding the laws of nature, we begin to understand ourselves.

Matthew Meyer
Author of Foundations of Physics: A Comprehensive Journey Through the Laws of the Universe

Foreword

Physics is more than a science — it is the foundation upon which every human achievement rests. From the sparks that power our cities to the stars that guide our exploration, physics reveals the hidden order that governs all things. It is the quiet architect of progress, the language through which the universe speaks, and the compass that points humanity toward understanding, innovation, and enlightenment.

Every advance in civilization — from the discovery of fire to the birth of quantum computing — has been shaped by our growing comprehension of nature's laws. When we learned to harness motion, we built engines. When we understood light, we created electricity, lasers, and communication across worlds. And when we began to probe the quantum and the cosmic, we glimpsed the very boundaries of reality itself.

To study physics is to study the story of existence — to ask not only *how* the universe works, but *why it works so beautifully*. It challenges us to think deeply, to reason clearly, and to see beyond appearances into the elegant simplicity beneath complexity. Physics trains the mind to seek truth through logic and evidence, yet also invites the heart to marvel at the harmony of creation.

In an age defined by technology and uncertainty, understanding the principles of physics is more vital than ever. It empowers us to build responsibly, to innovate wisely, and to see our place in the cosmos with humility and purpose. Each discovery, no matter how small, adds a new verse to humanity's ongoing dialogue with the universe.

Let this book serve as both a guide and an inspiration — a bridge between knowledge and wonder. For in exploring the laws of nature, we are not merely learning about the universe; we are learning about ourselves — our origins, our potential, and the infinite horizons of the human spirit.

Matthew Meyer
Foreword to Foundations of Physics: A Comprehensive Journey Through the Laws of the Universe

Part I – The Nature of Physics

Chapter 1: The Nature of Physics

Physics is the science of reality itself — the study of matter, energy, space, and time, and the relationships that govern them. It is both a practical discipline and a profound philosophy. From the falling of an apple to the bending of starlight, physics seeks to uncover the laws that make the universe coherent, measurable, and astonishingly beautiful.

At its heart, physics asks two questions:
"What is the universe made of?" and **"How does it work?"**
The answers to these questions have guided humanity's greatest discoveries and deepest reflections. Through physics, we've built bridges and spacecraft, discovered electricity and quantum particles, and learned that space and time are woven together into the fabric of existence itself.

1.1 The Spirit of Inquiry

Physics begins with curiosity. Ancient philosophers once wondered why the stars move in the heavens, why objects fall, and what makes fire burn. Their curiosity evolved into experimentation and mathematics — the tools that transformed wonder into understanding.

Today, physics remains a journey of exploration. We no longer study the cosmos by sight alone but by building machines that can peer into atoms and detect the faint echoes of the Big Bang. Yet the same question persists through millennia: *Why does the universe behave as it does?*

Every equation, every theory, and every observation is a response to that eternal curiosity.

1.2 The Scope of Physics

Physics is often divided into **branches**, each focusing on different aspects of nature:

- **Classical Mechanics** explains motion, forces, and energy — from rolling wheels to orbiting planets.

- **Thermodynamics** explores heat, temperature, and the laws governing energy transfer.

- **Electromagnetism** unites electricity, magnetism, and light under a single framework.

- **Relativity** redefines space and time, revealing how gravity is the geometry of spacetime.

- **Quantum Mechanics** unveils the strange, probabilistic behavior of particles at the smallest scales.

- **Nuclear and Particle Physics** probe the inner structure of atoms and the forces binding them.

- **Astrophysics and Cosmology** examine the stars, galaxies, and the birth and fate of the universe itself.

Together, these fields form a seamless continuum of understanding — from the infinitesimal to the infinite.

1.3 The Language of Nature: Mathematics

Mathematics is the grammar of the universe. Every physical law can be expressed as a mathematical relationship — concise, elegant, and universal. From Newton's $F = ma$ to Einstein's $E = mc^2$, equations allow us to describe how nature behaves with precision far beyond what words alone can capture.

In this way, mathematics and physics are inseparable. Mathematics gives structure to physical thought, and physics gives meaning to mathematical symbols. When we study physics, we are not merely solving problems; we are learning to interpret the poetry of existence written in numbers and patterns.

1.4 The Scientific Method

Physics thrives on **observation, experimentation, and reasoning**. The scientific method is its guiding principle — a disciplined way to transform curiosity into knowledge.

1. **Observe:** Notice a phenomenon or ask a question.

2. **Hypothesize:** Propose an explanation or model.

3. **Experiment:** Test the hypothesis through measurable outcomes.

4. **Analyze:** Compare predictions to results.

5. **Conclude:** Accept, refine, or reject the hypothesis.

6. **Repeat:** Continue exploring, questioning, and improving.

This process has led to every major advance in physics — from Galileo's falling spheres to the discovery of the Higgs boson. It reminds us that truth in science is not dictated but demonstrated.

1.5 Measurement, Units, and Precision

In physics, every observation must be **quantified**. We describe nature using **units of measurement**, which form the foundation of scientific accuracy. The modern standard system is the **SI (Système International)**, which includes:

- **Meter (m)** for length
- **Kilogram (kg)** for mass
- **Second (s)** for time
- **Ampere (A)** for electric current
- **Kelvin (K)** for temperature
- **Mole (mol)** for amount of substance
- **Candela (cd)** for luminous intensity

Precise measurement allows us to test theories, compare results, and communicate discoveries universally.

Physics values not only accuracy but also **uncertainty** —
understanding how reliable a measurement is.

1.6 The Unity of Physical Laws

One of the most profound insights of physics is
universality: the same laws apply everywhere. The force
that pulls an apple to the ground is the same force that
binds planets to the Sun. The equations that describe
electrons in an atom also describe the light from distant
galaxies.

This unity suggests that the cosmos, despite its vast
diversity, is governed by simple, elegant principles. To
uncover those principles is to glimpse the architecture of
reality itself.

1.7 The Human Journey Through Physics

Physics is not just a study of objects and forces — it is a
reflection of humanity's desire to know. Every discovery
tells a story of perseverance and imagination: Galileo
defying convention to study motion; Newton unveiling the
laws of gravity; Einstein redefining space and time;
modern scientists decoding the quantum world.

Each generation stands on the shoulders of the last,
continuing the quest for understanding. Physics is thus a
collective human endeavor — one that transcends
culture, time, and borders.

1.8 Why Physics Matters

Physics shapes every aspect of our world:

- It powers our technology — from smartphones to satellites.

- It fuels our industries — from renewable energy to medicine.

- It expands our vision — from quantum computation to interstellar travel.

But beyond utility, physics teaches us humility and wonder. It reminds us that the universe is vast, interconnected, and full of mysteries yet to be discovered. In learning physics, we do not diminish beauty by explaining it; we deepen it by understanding its truth.

1.9 The Journey Ahead

In the chapters that follow, we will explore the great domains of physics — beginning with the motion of objects and the forces that act upon them, then journeying through energy, light, relativity, and the fabric of the cosmos. Each part builds upon the last, forming a cohesive path toward understanding how nature works at every scale.

By the end of this journey, you will not only grasp the principles that govern the universe — you will see the universe as a vast, interconnected symphony of laws and patterns. Physics is not merely a subject to learn; it is a way of seeing, thinking, and being in harmony with reality itself.

Chapter 2: The Building Blocks of Reality

All of existence — from the light of distant stars to the atoms in our bodies — is composed of the same fundamental ingredients. Physics seeks to describe these ingredients, the forces that act upon them, and the laws that unify their behavior across every scale of the universe. In this chapter, we explore the foundations upon which all physical reality rests: **matter, energy, forces, particles, and fields**.

At its core, the universe is not a collection of separate things, but a dynamic, interconnected whole. What appears solid is a dance of invisible particles; what seems empty space is alive with fluctuating fields. The study of these fundamental structures reveals that everything we see, touch, and feel is built from a common cosmic code.

2.1 Matter and Energy

Matter is anything that has mass and occupies space. It forms the tangible universe — the planets, air, oceans, and living beings. At the atomic level, matter is composed of **atoms**, which themselves are made of **protons**, **neutrons**, and **electrons**.

Yet, matter is only half the story. The other half is **energy**, the capacity to cause change or do work. Energy exists in many forms — **kinetic, potential, thermal, electromagnetic, chemical,** and **nuclear** — but all can transform from one form to another without ever being destroyed. This principle is captured by the **Law of Conservation of Energy.**

Einstein's famous equation

$$E = mc^2$$

revealed the deep unity between matter and energy: they are two sides of the same coin. Mass can be converted into energy, as in nuclear reactions, and energy can manifest as matter, as seen in high-energy collisions. The universe itself is an ongoing transformation between these two states.

Thus, everything — from light waves to solid rock — is part of an eternal interplay between **mass and energy,** woven into the same cosmic fabric.

2.2 The Fundamental Forces of Nature

The motion and structure of the universe are governed by **four fundamental forces.** These forces determine how particles interact, how atoms form, how stars shine, and how galaxies hold together.

1. **Gravity**
   - The weakest but most far-reaching of the four forces.

- It acts between any objects with mass and is always attractive.

- Described by Newton's Law of Universal Gravitation and refined by Einstein's General Relativity, gravity curves space and time itself.

- It shapes the orbits of planets, the birth of stars, and the structure of the cosmos.

2. **Electromagnetism**

- Acts between charged particles.

- Responsible for electricity, magnetism, and light.

- Described by **Maxwell's Equations**, which unify electric and magnetic fields into one elegant framework.

- Electromagnetic forces bind atoms and molecules, making chemistry — and therefore life — possible.

3. **Strong Nuclear Force**

- The most powerful force in nature.

- It binds protons and neutrons together inside the atomic nucleus, overcoming the electromagnetic repulsion between positively charged protons.

- Carried by particles called **gluons**, it operates over subatomic distances — about the size of a proton.

- Without it, no atoms, stars, or matter as we know it could exist.

4. **Weak Nuclear Force**

- Responsible for radioactive decay and nuclear fusion in stars.

- It allows one type of particle to transform into another (for example, a neutron into a proton).

- Though weaker than electromagnetism and the strong force, it plays a vital role in powering the Sun and creating elements in the universe.

Together, these four forces form the foundation of all interactions in the cosmos. Physicists continue to seek a deeper unification — a theory that merges all four into one **"Theory of Everything."** The search for such a theory is one of science's greatest adventures.

2.3 Particles and Fields

In the traditional view, the universe was built from **particles** — tiny, solid bits of matter. But modern physics tells a subtler story: every particle is also a vibration or disturbance in an invisible **field** that permeates space.

A **field** is a quantity defined at every point in space and time.

- The **electric field** describes how a charge influences its surroundings.

- The **gravitational field** describes how mass curves space and time.

- The **quantum fields** describe the behavior of particles themselves.

In quantum field theory, what we call a "particle" is actually an **excitation of a field** — a localized ripple in an otherwise invisible ocean of energy.
For example:

- An **electron** is a vibration of the **electron field**.

- A **photon** is a vibration of the **electromagnetic field**.

- A **quark** is a vibration of the **quark field**.

This view unifies matter and forces in a single framework: fields interacting with other fields. The universe, then, is not made of "things" but of **interactions and relationships** — a continuous symphony of fields exchanging energy.

2.4 The Standard Model Overview

The **Standard Model of Particle Physics** is one of humanity's most successful scientific achievements. It describes all known fundamental particles and the forces (except gravity) that govern them.

It classifies the building blocks of matter into two main families:

1. Fermions — The Particles of Matter

These make up everything that has substance.

- **Quarks:** Combine to form protons and neutrons (up, down, charm, strange, top, bottom).

- **Leptons:** Include the electron, muon, tau, and their associated neutrinos.

2. Bosons — The Particles of Force

These carry the fundamental interactions.

- **Photon (γ):** Carries electromagnetism.

- **Gluons (g):** Carry the strong nuclear force.

- **W and Z Bosons:** Carry the weak nuclear force.

- **Higgs Boson (H):** Gives mass to other particles through the Higgs field.

Although the **graviton** (the theoretical particle of gravity) has not yet been discovered, it remains part of the quest for a unified theory that joins **quantum mechanics** and **general relativity**.

The Standard Model has been experimentally verified to extraordinary precision. Yet, it leaves profound questions unanswered:

- Why do particles have the masses they do?

- What is dark matter and dark energy?

- Why is there more matter than antimatter?

These mysteries suggest that the Standard Model, while powerful, is only a chapter in a much larger cosmic story.

2.5 The Scale of the Universe — From Quarks to Galaxies

The universe spans an almost unimaginable range of scales — from the **infinitesimal** to the **infinite**.

Scale	Approximate Size	Description
Quarks	10^{-18} meters	The smallest known constituents of matter.
Protons/Neutrons	10^{-15} meters	The building blocks of atomic nuclei.
Atoms	10^{-10} meters	The basic units of matter, forming molecules.
Cells	10^{-6} meters	The building blocks of life.

Scale	Approximate Size	Description
Humans	10^0 meters	Observers of the universe's grandeur.
Planets	10^7 meters	Homes of life and motion.
Stars	10^9 meters	Cosmic furnaces that create the elements.
Galaxies	10^{21} meters	Vast collections of stars bound by gravity.
Observable Universe	10^{26} meters	The total reach of light since the Big Bang.

From quarks to galaxies, the same physical principles govern every level of reality. The gravitational attraction that shapes galaxies is described by the same universal constants that define atomic behavior. The universe is a continuum — a seamless hierarchy of scales bound together by the language of physics.

2.6 A Universe of Unity

Though it may seem diverse and chaotic, the universe is unified by a profound simplicity. Four forces, a handful of particles, and a few fundamental constants give rise to everything that exists — stars, oceans, mountains, and the human mind itself.

Physics reveals that we are not separate from the cosmos, but expressions of it — intricate arrangements of the same energy that fuels the stars. To study the building blocks of reality is to understand the architecture of creation, and our place within it.

Part II – Classical Mechanics: The Science of Motion

Chapter 3: Motion and Kinematics

Motion is the most fundamental expression of change in the universe. From the spinning of electrons to the orbits of galaxies, everything moves, shifts, and evolves through time. Understanding motion is the first step in understanding the forces and energies that shape our world.

Kinematics, the branch of mechanics that describes motion without regard to its causes, provides the language we use to measure and predict how things move. It answers the questions: *Where is an object? How fast is it moving? How is its motion changing?*

This chapter introduces the essential quantities — **position, velocity, and acceleration** — and the mathematical tools that allow us to describe and visualize motion across space and time.

3.1 Position, Velocity, and Acceleration

To study motion, we first define where an object is and how that position changes.

Position

An object's **position** describes its location relative to a chosen reference point, known as the **origin**.
In one dimension, position is represented by a coordinate x along a straight line.

In two or three dimensions, we use coordinates (x, y) or (x, y, z).

- **Displacement (Δx)** is the change in position:

$$\Delta x = x_f - x_i$$

where x_i and x_f are the initial and final positions. Displacement differs from *distance* — it has direction as well as magnitude.

Velocity

Velocity measures how fast and in what direction an object's position changes over time.

- **Average velocity:**

$$v_{avg} = \frac{\Delta x}{\Delta t}$$

- **Instantaneous velocity:**
  The velocity at a specific moment, found using calculus:

$$v = \frac{dx}{dt}$$

Velocity is a **vector quantity**, meaning it includes both magnitude (speed) and direction.

Acceleration

Acceleration is the rate at which velocity changes over time.

It tells us how quickly an object speeds up, slows down, or changes direction.

- **Average acceleration:**

$$a_{avg} = \frac{\Delta v}{\Delta t}$$

- **Instantaneous acceleration:**

$$a = \frac{dv}{dt}$$

Acceleration is also a **vector**, pointing in the direction of the change in velocity.

If velocity and acceleration point in the same direction, the object speeds up. If they point in opposite directions, the object slows down.

3.2 Graphs of Motion

Graphs provide a powerful visual tool to represent motion. By plotting quantities such as position, velocity, and acceleration against time, we can understand how an object's motion evolves at a glance.

Position–Time Graphs

- The **slope** of a position–time graph gives **velocity**. A straight line indicates **constant velocity**.

A curved line shows **changing velocity (acceleration)**.

- The steeper the slope, the faster the motion.

Velocity–Time Graphs

- The **slope** of a velocity–time graph gives **acceleration**.

- The **area under the curve** represents **displacement**.
  For constant acceleration, this area forms a simple geometric shape, such as a rectangle or triangle.

Acceleration–Time Graphs

- The **area under the acceleration–time curve** gives the **change in velocity**.

- A horizontal line indicates constant acceleration; a line along zero shows constant velocity.

By interpreting these graphs, we can translate between the visual and mathematical forms of motion — a skill essential for all physicists and engineers.

3.3 Free Fall and Projectile Motion

Free Fall

When an object falls under the influence of gravity alone, it is in **free fall**.

Near Earth's surface, the acceleration due to gravity is approximately:

$$g = 9.8 \text{ m/s}^2$$

Key equations for uniformly accelerated motion (assuming upward is positive and air resistance is negligible):

$$v = v_0 + at$$
$$x = x_0 + v_0 t + \frac{1}{2}at^2$$
$$v^2 = v_0^2 + 2a(x - x_0)$$

For free fall, $a = -g$.

These equations describe the rise and fall of objects — from tossed stones to launched rockets.

Projectile Motion

A **projectile** is any object that moves through the air under the influence of gravity alone.

Its motion combines two independent components:

- **Horizontal motion:** Constant velocity, since there is no horizontal acceleration (neglecting air resistance).

- **Vertical motion:** Accelerated motion due to gravity.

This combination creates a **parabolic path**.

Equations of motion:

$$x = v_0 \cos(\theta)\, t$$
$$y = v_0 \sin(\theta)\, t - \frac{1}{2} g t^2$$

Where v_0 is the initial speed and θ the launch angle.
At $\theta = 45°$, the projectile achieves **maximum range** (in ideal conditions).

Projectiles illustrate how gravity influences all motion — whether it's a cannonball, a football, or a satellite in orbit.

3.4 Relative Motion

All motion is **relative** — it depends on the observer's frame of reference.
There is no absolute rest in the universe; everything moves relative to something else.

If a car moves east at 60 km/h and a passenger walks forward inside it at 5 km/h (relative to the car), then relative to the ground, the passenger moves at:

$$v_{total} = v_{car} + v_{passenger} = 65 \text{ km/h}$$

But to another car moving alongside at the same speed, both appear motionless relative to each other.

This principle — the **relativity of motion** — extends beyond everyday experience. It laid the groundwork for Einstein's **Special Theory of Relativity**, which revealed that time and space themselves are relative quantities.

In classical physics, however, we assume **Galilean relativity**, where the laws of motion are the same in all inertial (non-accelerating) frames of reference.

3.5 The Language of Motion

Kinematics gives us a precise language to describe the movement of everything — from a falling leaf to a planet in orbit.
It lays the foundation for **dynamics**, where we explore the forces that cause motion.

By mastering position, velocity, acceleration, and the tools to visualize them, we gain insight into one of nature's simplest yet most profound truths: **motion connects every part of the universe**.

From the steady drift of galaxies to the flutter of an atom, everything in creation is in motion — an eternal rhythm that defines the living universe.

Chapter 4: Forces and Newton's Laws of Motion

All motion arises from interaction. When a ball rolls, a planet orbits, or a rocket ascends, unseen influences act upon them — pushing, pulling, and guiding their paths. These influences are called **forces**, and they are the language through which nature expresses change.

This chapter introduces **Isaac Newton's Three Laws of Motion**, which describe how forces affect the motion of objects. These laws form the cornerstone of classical mechanics and remain among the most profound achievements in human thought. They reveal that motion is not random but governed by precise, universal principles.

4.1 Newton's Three Laws of Motion

In 1687, Sir Isaac Newton published *Philosophiæ Naturalis Principia Mathematica*, laying the foundation for modern physics. His three laws describe the relationship between motion and force and apply to all objects, from falling apples to orbiting moons.

Newton's First Law – The Law of Inertia

"An object at rest will remain at rest, and an object in motion will continue in motion at a constant velocity, unless acted upon by a net external force."

This law defines **inertia** — the natural tendency of an object to resist changes in its state of motion.
In the absence of an external force, an object will not accelerate. This means:

- A ball rolling on a frictionless surface would roll forever.

- A stationary object will not move unless something acts upon it.

Inertia depends on **mass** — the greater an object's mass, the greater its resistance to change in motion.

Newton's Second Law – The Law of Acceleration

"The acceleration of an object is directly proportional to the net force acting on it and inversely proportional to its mass."

This law provides the mathematical foundation of motion:

$$F = ma$$

Where:

- F is the **net force** (in newtons, N),

- m is the **mass** (in kilograms, kg),

- a is the **acceleration** (in meters per second squared, m/s^2).

This equation means:

- The greater the force, the greater the acceleration.

- The greater the mass, the smaller the acceleration for the same force.

This principle governs everything from the motion of cars to the flight of rockets.
It also introduces the **newton (N)** — the SI unit of force:

$$1N = 1kg \cdot m/s2$$

Newton's Third Law – The Law of Action and Reaction

"For every action, there is an equal and opposite reaction."

Whenever one object exerts a force on another, the second object exerts an equal force in the opposite direction.
These forces occur **in pairs** and act on **different objects**.

Examples:

- When you push a wall, it pushes back with equal force.

- A rocket's exhaust gases push downward, and the rocket is propelled upward.

- The ground pushes back on your feet as you walk, allowing you to move forward.

The third law illustrates that forces are interactions — they exist only in relationships between objects.

4.2 Inertia, Force, and Acceleration

The concepts of **inertia, force**, and **acceleration** are deeply interconnected.

- **Inertia** describes an object's resistance to acceleration.

- **Force** is what overcomes inertia to produce acceleration.

- **Acceleration** is the observable result of an unbalanced (net) force acting on mass.

Net Force and Equilibrium

If the sum of all forces acting on an object is zero ($\Sigma F = 0$), the object is in **equilibrium** — it either remains at rest or moves with constant velocity.
If there is a **net force**, the object accelerates in the direction of that force.

Free-Body Diagrams

Physicists use **free-body diagrams** to visualize all forces acting on an object.
Each force is drawn as a vector (arrow), representing both magnitude and direction.
Analyzing these forces allows us to apply Newton's laws quantitatively.

4.3 Friction, Tension, and Normal Forces

Forces in the real world often involve contact — surfaces pushing, pulling, or resisting motion. Three of the most common are **friction, tension,** and **normal force.**

Frictional Force (f)

Friction opposes motion between two surfaces in contact.
It arises from microscopic irregularities and electromagnetic interactions between surfaces.

- **Static friction (f_s):** The force that prevents motion.

$$f_s \leq \mu_s N$$

- **Kinetic friction (f_k):** The force resisting motion once sliding occurs.

$$f_k = \mu_k N$$

Here, μ_s and μ_k are the **coefficients of friction**, and N is the **normal force**.

Friction converts kinetic energy into heat, slowing objects down — yet it's also essential for walking, driving, and holding objects in place.

Tension Force (T)

Tension is the pulling force transmitted through a rope, string, or cable when it is taut.
It acts along the length of the medium and always pulls

away from the object.
If two masses are connected by a rope over a pulley, the tension acts equally on both (neglecting friction and mass of the rope).

Tension provides the mechanical link in systems ranging from elevators to bridges and pendulums.

Normal Force (N)

The **normal force** is the support force exerted by a surface perpendicular to an object resting on it.
If you place a book on a table, gravity pulls downward, but the table pushes upward with equal magnitude — the normal force.

For a flat surface:

$$N = mg$$

On an inclined plane:

$$N = mg\cos(\theta)$$

The normal force prevents objects from "falling through" surfaces and is essential for balancing forces in static and dynamic systems.

4.4 Circular Motion and Centripetal Force

When an object moves in a circle at constant speed, its direction of motion is continually changing — meaning it is constantly **accelerating**, even if its speed remains constant. This acceleration is called **centripetal acceleration**, and it always points toward the **center** of the circular path.

$$a_c = \frac{v^2}{r}$$

To cause this acceleration, a **centripetal force** must act toward the center of the circle:

$$F_c = m\frac{v^2}{r}$$

This force is not a new kind of force; it can be provided by **gravity**, **tension**, **friction**, or **normal forces**, depending on the situation.

Examples:

- The tension in a string keeps a swinging ball moving in a circle.

- Friction between tires and road keeps a car turning.

- Gravity keeps planets orbiting the Sun.

If the centripetal force disappears, the object moves **tangentially** to the circle — in a straight line, as Newton's First Law predicts.

4.5 The Harmony of Motion and Force

Newton's laws reveal that motion and force are inseparable.
Every change in motion has a cause; every force creates an equal and opposite response.

From the smallest grain of dust to entire galaxies, these same laws hold true — timeless, universal, and exact. They are the framework upon which all of classical mechanics is built, and they remain the foundation of physics today.

Newton once said, *"If I have seen further, it is by standing on the shoulders of giants."*
His own discoveries became the shoulders upon which modern physics stands — guiding us toward the stars, into the atom, and beyond.

Chapter 5: Work, Energy, and Power

Motion is not only governed by forces — it is sustained and transformed through **energy**. Every falling apple, spinning planet, or glowing star owes its behavior to the constant conversion of energy from one form to another.

Work, energy, and **power** are fundamental ideas that bridge the gap between dynamics and the deeper understanding of how the universe evolves. This chapter explores how forces produce work, how energy manifests and transforms, and how the total energy of an isolated system remains constant — a profound principle known as the **Law of Conservation of Energy**.

5.1 Work Done by a Force

Whenever a **force** causes an object to move, **work** is done. Work measures how much energy is transferred by a force acting through a distance.

Definition of Work

In physics, **work (W)** is defined as:

$$W = Fd\cos(\theta)$$

Where:

- W = work (in joules, J)

- F = magnitude of the applied force (in newtons, N)

- d = displacement of the object (in meters, m)
- θ = angle between the force and displacement vectors

Key Principles

- Only the **component of the force parallel** to the displacement does work.

- If $\theta = 0°$, all the force contributes to work ($W = Fd$).

- If $\theta = 90°$, the force does **no work** (e.g., a centripetal force in circular motion).

- **Positive work** increases the object's energy (force in direction of motion).

- **Negative work** decreases energy (force opposite to motion, like friction).

Units of Work

The SI unit of work is the **joule (J)**:

$$1J = 1N \cdot m = 1kg \cdot m2/s2$$

One joule represents the work done by a force of one newton acting over a distance of one meter.

Work–Energy Connection

When a force does work on an object, it changes the object's **energy**.

This relationship is expressed in the **Work–Energy Theorem**:

$$W = \Delta K$$

The net work done by all forces equals the change in **kinetic energy** of the object.

5.2 Kinetic and Potential Energy

Energy is the capacity to do work. It exists in many forms, but two are most central to mechanics: **kinetic** and **potential** energy.

Kinetic Energy (K)

Kinetic energy is the energy of motion.
Any moving object — from a drifting leaf to a speeding comet — possesses kinetic energy given by:

$$K = \frac{1}{2}mv^2$$

Where:

- m = mass (kg)

- v = velocity (m/s)

Because velocity is squared, doubling an object's speed **quadruples** its kinetic energy.

Kinetic energy is always **positive** and depends entirely on motion.

Potential Energy (U)

Potential energy is **stored energy** — energy due to an object's position, configuration, or condition. It represents the potential to do work.

Gravitational Potential Energy

An object lifted above the ground has gravitational potential energy due to Earth's gravitational field:

$$U_g = mgh$$

Where:

- m= mass (kg)
- g= acceleration due to gravity (9.8 m/s^2)
- h= height (m) above a reference level

If the object falls, this stored energy converts to kinetic energy.

Elastic Potential Energy

Objects like springs or rubber bands store energy when stretched or compressed:

$$U_s = \frac{1}{2}kx^2$$

Where:

- k = spring constant (N/m)

- x = displacement from equilibrium (m)

The greater the stretch, the greater the stored energy — as long as the elastic limit is not exceeded.

Energy Transformation

Energy constantly changes form — from potential to kinetic, mechanical to thermal, or chemical to electrical. When you drop a ball:

- Gravitational potential energy decreases.

- Kinetic energy increases.

- The total energy remains constant.

This interplay underlies every process in the physical universe.

5.3 Conservation of Energy

The **Law of Conservation of Energy** states that **energy cannot be created or destroyed — only transformed from one form to another**.
In a closed system, the total energy remains constant over time.

$$E_{total} = K + U = \text{constant}$$

In other words:

$$\Delta K + \Delta U = 0$$

When no external forces (like friction) do work, the sum of kinetic and potential energy — the **mechanical energy** — is conserved.

Examples of Energy Conservation

1. **Free Fall:**
   At the top of the fall, energy is all potential ($U = mgh$).
   At the bottom, it is all kinetic ($K = \frac{1}{2}mv^2$).
   The total energy remains the same at every point.

2. **Pendulum:**
   As the pendulum swings, potential energy converts to kinetic and back again — perfectly exchanging between forms in an ideal (frictionless) system.

3. **Roller Coaster:**
   Each hill, drop, and loop transforms energy between potential and kinetic, while some is lost as heat and sound due to friction — showing how real-world systems are not perfectly isolated.

Non-Conservative Forces

In real systems, forces like **friction** and **air resistance** convert mechanical energy into heat, sound, or deformation.

Though mechanical energy decreases, **total energy (including thermal)** is still conserved — the universe keeps perfect accounting.

5.4 Power and Efficiency

Power

Power measures **how quickly work is done** or **how fast energy is transferred**.

$$P = \frac{W}{t}$$

Where:

- P = power (in watts, W)
- W = work done (in joules, J)
- t = time (in seconds, s)

Another useful form links power to force and velocity:

$$P = Fv$$

Units of Power

The SI unit of power is the **watt (W)**:

$$1\,W = 1\,J/s$$

Larger units include the **kilowatt (kW)** and **horsepower (hp)**, where:

$$1 \text{ hp} = 746 \text{ W}$$

Power tells us how fast energy changes form — a 100 W lightbulb converts 100 joules of electrical energy into light and heat every second.

Efficiency

Not all energy input becomes useful output. **Efficiency** measures how effectively energy is converted from one form to another.

$$\text{Efficiency} = \frac{\text{Useful Output Energy}}{\text{Total Input Energy}} \times 100\%$$

No process is 100% efficient — some energy is always lost as heat or friction.
A car engine, for example, converts only about 25–30% of its fuel's energy into motion; the rest dissipates as heat.

Efficiency reminds us that while **energy is conserved**, its *usefulness* can degrade — a concept tied to **entropy** in thermodynamics.

5.5 The Continuum of Work, Energy, and Power

Work connects **force** and **motion**.
Energy describes **the capacity to perform work**.
Power expresses **the rate at which work or energy is used**.

Together, they form the foundation for understanding everything from engines and orbits to stars and galaxies.

In the smallest scales, they describe the motion of particles; in the largest, the evolution of the cosmos. Through them, we see that the universe is not static — it is dynamic, ever-flowing, and ever-transforming, governed by the eternal conservation of energy.

Chapter 6: Momentum and Collisions

In the dynamic universe, motion is rarely isolated. Objects interact, collide, and exchange energy and motion in endlessly intricate ways. To describe these interactions, physicists use one of the most enduring and powerful concepts in science: **momentum**.

Momentum measures how difficult it is to stop a moving object — its "quantity of motion." It is deeply tied to both mass and velocity, and it obeys one of the most profound laws in physics: **the Law of Conservation of Momentum**.

This chapter explores momentum, its connection to force and impulse, and how it governs collisions, explosions, and rocket motion — from the simplest impacts to the grandest cosmic events.

6.1 Linear Momentum and Impulse

Definition of Linear Momentum

The **linear momentum (p)** of an object is the product of its mass and velocity:

$$\mathbf{p} = m\mathbf{v}$$

Where:

- **p**= momentum (in kg·m/s)

- m= mass (in kg)

- **v**= velocity (in m/s)

Momentum is a **vector quantity** — it has both magnitude and direction.
The greater an object's mass or speed, the greater its momentum and the harder it is to change its motion.

Impulse and Change in Momentum

A force applied over a period of time changes an object's momentum. This change is called **impulse (J)**.

$$J = F\Delta t = \Delta p$$

Where:

- **J**= impulse (in N·s)

- **F**= average force applied

- Δt= time interval

- **Δp**= change in momentum

Impulse equals the change in momentum.

This relationship is known as the **Impulse–Momentum Theorem**.

It tells us that a small force applied over a long time can produce the same change in momentum as a large force applied briefly. For example:

- A baseball bat hitting a ball exerts a large force for a short time.

- A car's airbag exerts a smaller force over a longer time, reducing injury while achieving the same momentum change.

Units of Momentum and Impulse

Momentum:

$$kg \cdot m/s$$

Impulse:

$$N \cdot s = kg \cdot m/s$$

6.2 Elastic and Inelastic Collisions

Collisions occur when two or more objects exert forces on each other for a short duration.
All collisions obey the **Law of Conservation of Momentum**, but they differ in how **kinetic energy** is conserved or transformed.

Elastic Collisions

In an **elastic collision**, both **momentum** and **kinetic energy** are conserved.

$$m_1 v_{1i} + m_2 v_{2i} = m_1 v_{1f} + m_2 v_{2f}$$
$$\{\frac{1}{2} m_1 v_{1i}^2 + \frac{1}{2} m_2 v_{2i}^2 = \frac{1}{2} m_1 v_{1f}^2 + \frac{1}{2} m_2 v_{2f}^2$$

Examples:

- Collisions between billiard balls

- Gas molecules colliding in a container

- Atomic and subatomic particle scattering

In an ideal elastic collision, no kinetic energy is lost to heat, sound, or deformation.

Inelastic Collisions

In an **inelastic collision, momentum is conserved**, but **kinetic energy is not**.
Some energy is transformed into heat, sound, or internal deformation.

The most extreme case is a **perfectly inelastic collision**, where objects stick together after impact.

$$m_1 v_{1i} + m_2 v_{2i} = (m_1 + m_2) v_f$$

Although kinetic energy decreases, total energy (including heat and deformation) remains conserved — in accordance with the **Law of Conservation of Energy**.

Comparing Collisions

Type of Collision	Momentum Conserved	Kinetic Energy Conserved	Example
Elastic	☑ Yes	☑ Yes	Billiard balls, atomic collisions
Inelastic	☑ Yes	✗ No	Car crash, clay ball collision
Perfectly Inelastic	☑ Yes	✗ No	Objects stick together

In reality, most collisions are **partially inelastic**, with some energy lost but not all.

6.3 Conservation of Momentum

The **Law of Conservation of Momentum** states:

In a closed, isolated system (no external forces), the total momentum before interaction equals the total momentum after interaction.

Mathematically:

$$\sum \mathbf{p}_{\text{initial}} = \sum \mathbf{p}_{\text{final}}$$

or, for two objects:

$$m_1 v_{1i} + m_2 v_{2i} = m_1 v_{1f} + m_2 v_{2f}$$

This principle is one of the most fundamental in all of physics. It applies universally — to atoms in a gas, vehicles on a highway, and even galaxies in collision.

Momentum conservation arises directly from a deep symmetry in nature: **the uniformity of space**. Because the laws of physics are the same everywhere, momentum must be conserved — a concept formalized in **Noether's Theorem**.

Examples of Momentum Conservation

1. **Recoil of a Gun**
   When a gun fires a bullet, the bullet moves forward, and the gun recoils backward. The total momentum before and after firing remains zero.

$$m_{gun} v_{gun} = -m_{bullet} v_{bullet}$$

2. **Colliding Carts**
   When two carts collide on a frictionless track, their combined momentum remains constant, even if they exchange velocities.

3. **Explosion of a Firework**
   The fragments fly in opposite directions with equal and opposite momenta, keeping the total momentum of the system unchanged.

6.4 Rocket Propulsion

Perhaps the most striking application of momentum conservation is **rocket propulsion** — a triumph of Newton's Third Law in action.

A rocket moves not by pushing against the air, but by **expelling mass** (exhaust gases) backward.
The backward momentum of the exhaust equals the forward momentum gained by the rocket.

Principle of Conservation of Momentum in Rockets

$$m_1 v_{1i} + m_2 v_{2i} = m_1 v_{1f} + m_2 v_{2f}$$

Initially, the rocket and fuel are at rest. As fuel burns and gas is expelled backward at high speed, the rocket moves forward with equal and opposite momentum.

Rocket Equation

Derived from momentum conservation, the **Tsiolkovsky Rocket Equation** expresses how a rocket's velocity changes as it burns fuel:

$$\Delta v = v_e \ln \left(\frac{m_i}{m_f} \right)$$

Where:

- Δv = change in velocity

- v_e = exhaust velocity

- m_i = initial mass (with fuel)

- m_f = final mass (without fuel)

This equation shows that the faster the exhaust gases and the greater the mass of expelled fuel, the higher the rocket's velocity.
It also explains why multi-stage rockets are needed for space travel — shedding mass dramatically improves efficiency.

6.5 The Eternal Balance of Motion

Momentum reveals the universe's perfect bookkeeping. In every collision, explosion, or recoil, the total momentum — the total *motion of existence* — is always conserved.

While energy can change form and flow from one object to another, momentum remains an unbroken thread of continuity connecting every interaction.

From atoms colliding in a gas to galaxies merging in the depths of space, the law is the same:
Momentum is never lost, only shared.

It is one of nature's most elegant truths — a law that underlies all dynamics and ensures that, through every impact and transformation, the universe remains in balance.

Chapter 7: Rotational Motion and Gravity

Motion in the universe is rarely linear. From the spin of atoms to the orbits of planets, nature moves in circles, spirals, and rotations. Understanding rotational motion allows us to see the same laws that govern a spinning wheel also describe the motion of moons, stars, and galaxies.

This chapter explores **angular motion, torque, rotational dynamics**, and the law that unites all celestial movement — **Newton's Law of Universal Gravitation**. Together, these principles reveal the deep unity between motion on Earth and in the heavens.

7.1 Angular Displacement, Velocity, and Acceleration

When objects rotate, they move along circular paths about a fixed axis.
In **rotational motion**, we replace linear quantities with their **angular equivalents**.

Angular Displacement (θ)

Angular displacement measures how much an object has rotated, defined as the angle swept by a line joining the object to the axis of rotation.

$$\theta = \frac{s}{r}$$

Where:

- θ= angular displacement (in radians)
- s= arc length (m)
- r= radius of rotation (m)

One **radian** is the angle subtended when the arc length equals the radius.
A full revolution equals:

$$2\pi \text{ radians} = 360°$$

Angular Velocity (ω)

Angular velocity describes how fast an object rotates — the rate of change of angular displacement with time.

$$\omega = \frac{d\theta}{dt}$$

Where:

- ω= angular velocity (in radians per second, rad/s)

The **linear velocity** (v) of a point on the rotating object relates to its angular velocity by:

$$v = \omega r$$

This means all points on a rotating body share the same angular velocity, but points farther from the center move faster linearly.

Angular Acceleration (α)

Angular acceleration is the rate of change of angular velocity:

$$\alpha = \frac{d\omega}{dt}$$

Where:

- α= angular acceleration (in radians per second squared, rad/s^2)

Rotational motion with constant angular acceleration follows equations analogous to linear kinematics:

$$\omega_f = \omega_i + \alpha t$$
$$\theta = \omega_i t + \frac{1}{2}\alpha t^2$$
$$\omega_f^2 = \omega_i^2 + 2\alpha\theta$$

These equations describe spinning wheels, rotating machinery, and celestial rotation alike — unified by mathematical symmetry.

7.2 Torque and Rotational Dynamics

Just as force causes linear acceleration, **torque** causes **angular acceleration**. Torque measures how effectively a force can cause an object to rotate about an axis.

Definition of Torque (τ)

$$\tau = rF\sin(\theta)$$

Where:

- τ= torque (in newton-meters, N·m)
- r= lever arm distance from the axis of rotation (m)
- F= applied force (N)
- θ= angle between the force and lever arm

Torque depends on both the magnitude of the force and its distance from the pivot.
A force applied farther from the axis or at a right angle produces a larger torque.

Rotational Inertia (Moment of Inertia, I)

An object's resistance to changes in rotation is measured by its **moment of inertia**, the rotational analog of mass.

$$I = \sum mr^2$$

The farther the mass is distributed from the axis, the greater the moment of inertia.

For example:

- Solid disk: $I = \frac{1}{2}mr^2$

- Thin hoop: $I = mr^2$

- Solid sphere: $I = \frac{2}{5}mr^2$

Newton's Second Law for Rotation

The relationship between torque and angular acceleration is:

$$\tau_{net} = I\alpha$$

This is the **rotational form of Newton's Second Law**. It tells us that torque produces angular acceleration proportional to the applied torque and inversely proportional to rotational inertia.

Angular Momentum (L)

The rotational equivalent of linear momentum is **angular momentum**, defined as:

$$L = I\omega$$

Angular momentum is **conserved** when no external torque acts on a system:

$$L_i = L_f$$

This principle explains why:

- A spinning figure skater speeds up when pulling in their arms (reducing I).

- Planets move faster when closer to the Sun (Kepler's Second Law).

- The universe maintains rotational balance across cosmic scales.

7.3 Newton's Law of Universal Gravitation

Isaac Newton's greatest insight was recognizing that the same force that causes an apple to fall also keeps the Moon in orbit.

He formulated the **Law of Universal Gravitation**:

$$F = G\frac{m_1 m_2}{r^2}$$

Where:

- F= gravitational force between two masses (N)

- G= gravitational constant ($6.674 \times 10^{-11} N \cdot m^2/kg^2$)

- m_1, m_2= interacting masses (kg)

- r= distance between their centers (m)

This law shows that:

- Gravity acts between **all objects** with mass.

- It is always **attractive**.

- Its strength decreases with the **square of the distance** between objects.

Despite its weakness compared to other forces, gravity dominates at large scales because it is **always positive and cumulative**, binding planets, stars, and galaxies into cosmic harmony.

Weight as a Gravitational Force

On Earth's surface, the force of gravity acting on an object is its **weight**:

$$W = mg$$

Where $g = 9.8$ m/s^2.
Weight depends on location (gravitational field strength), but mass remains constant everywhere.

7.4 Orbits and Kepler's Laws

Before Newton, **Johannes Kepler** derived three laws describing planetary motion based on careful observations by Tycho Brahe. Newton later showed these laws are natural consequences of universal gravitation.

Kepler's First Law – The Law of Ellipses

Every planet moves in an elliptical orbit with the Sun at one focus.

An **ellipse** is an oval-shaped curve. The Sun is not at the center but at one of the two **foci**. This explains why planets sometimes move closer to or farther from the Sun.

Kepler's Second Law – The Law of Equal Areas

A line joining a planet and the Sun sweeps out equal areas in equal times.

Planets move faster when nearer to the Sun and slower when farther away.
This reflects the **conservation of angular momentum** — as the distance decreases, velocity must increase.

$$L = mvr = \text{constant}$$

Kepler's Third Law – The Law of Harmonies

The square of a planet's orbital period is proportional to the cube of its average distance from the Sun.

$$T^2 \propto r^3$$

or

$$\frac{T^2}{r^3} = \text{constant}$$

Where:

- T = orbital period (time to complete one orbit)

- r = average orbital radius

This law applies to all planets and satellites orbiting a massive body — whether the Sun, a planet, or a star. Newton later derived this relationship directly from his gravitational law, uniting terrestrial and celestial motion under the same physics.

7.5 The Unity of Rotation and Gravitation

Rotation and gravitation are two expressions of the same universal order.
From a spinning top to the spiral arms of a galaxy, every motion reflects the same interplay between mass, distance, and force.

Newton's revelation that the laws governing falling apples and orbiting moons are one and the same forever changed humanity's view of the cosmos. It showed that **the universe is not a collection of isolated events but a single, harmonious system governed by universal laws**.

Through the study of rotational motion and gravity, we glimpse the geometric beauty of creation — a universe where every orbit, every spin, and every motion participates in the grand celestial dance of physics.

Part III – Thermodynamics: The Science of Heat and Energy

Chapter 8: Temperature, Heat, and Thermal Energy

Everything in the universe is in motion — not only planets and stars, but the tiniest atoms and molecules. Even in stillness, matter vibrates, spins, and collides in a constant dance of energy. This invisible motion of particles gives rise to **temperature, heat**, and **thermal energy**, the central concepts of **thermodynamics** — the science of energy, work, and transformation.

Thermodynamics reveals that heat is not a mysterious fluid, as once believed, but a measurable form of energy transfer. Through its laws, we understand why engines run, why ice melts, why stars burn, and why time seems to flow in one direction.

8.1 Temperature and Its Scales

What Is Temperature?

Temperature is a measure of the **average kinetic energy** of the particles in a substance.
The faster the particles move, the higher the temperature; the slower they move, the lower it is. At absolute zero, motion nearly ceases — the ultimate cold limit of the universe.

Temperature is not the total amount of energy in a substance, but rather an **average energy per particle**. A large lake and a cup of hot tea can contain vastly different amounts of thermal energy even if they share the same temperature.

Temperature Scales

Over centuries, scientists have developed several temperature scales to quantify thermal states. The three most widely used are **Celsius, Fahrenheit**, and **Kelvin.**

Celsius (°C)

- Based on the behavior of water.
- Defined so that:
    - 0°C= freezing point of water
    - 100°C= boiling point of water (at 1 atmosphere)
- Commonly used in most of the world and in scientific contexts for everyday measurements.

Fahrenheit (°F)

- Developed by Daniel Gabriel Fahrenheit in 1724.
- Defined so that:
    - 32°F= freezing point of water
    - 212°F= boiling point of water

- Primarily used in the United States.

Conversion between Celsius and Fahrenheit:

$$T_F = \frac{9}{5} T_C + 32$$

$$T_C = \frac{5}{9} (T_F - 32)$$

Kelvin (K)

- The **absolute temperature scale**, used universally in science.

- Begins at **absolute zero (0 K)** — the point at which all molecular motion theoretically stops.

- Defined so that:

$$T_K = T_C + 273.15$$

- No degree symbol (°) is used with Kelvin.

- Absolute zero = $-273.15°$C= $-459.67°$F.

The Kelvin scale is fundamental to physics because it reflects the **true physical nature of temperature**, free from arbitrary reference points.

8.2 Heat and Thermal Energy

Thermal Energy

Thermal energy is the total internal energy of a system due to the motion of its particles.
It depends on:

- Temperature (average kinetic energy)

- Mass (number of particles)

- Type of substance (bond structure and degrees of freedom)

A hot cup of water and a cold ocean wave illustrate this distinction: the cup has higher temperature, but the ocean contains far more thermal energy due to its mass.

Heat

Heat (Q) is the **transfer of thermal energy** from one object to another because of a temperature difference. It always flows **spontaneously from hot to cold**, until thermal equilibrium is reached.

Heat is **not a substance** — it is **energy in motion**. When heat flows into a system, the internal energy increases; when it flows out, the internal energy decreases.

$Q > 0$ (heat added to system); $Q < 0$ (heat lost by system)

The SI unit of heat is the **joule (J)**, the same as for work and energy.
In older systems, heat was sometimes measured in **calories**, where:

$$1 \text{ cal} = 4.184 \text{ J}$$

8.3 Heat Transfer: Conduction, Convection, and Radiation

Energy in the form of heat can move in three ways — **conduction**, **convection**, and **radiation**. These mechanisms operate everywhere, from a stove's surface to the core of stars.

Conduction

Conduction is the transfer of heat **through direct contact**.

When a metal rod is heated at one end, its atoms vibrate more rapidly and collide with neighboring atoms, passing on energy.

- Occurs mainly in **solids**, especially metals (good conductors).

- Poor conductors (like wood or air) are called **insulators**.

The rate of heat transfer by conduction is given by **Fourier's Law**:

$$\frac{Q}{t} = kA\frac{\Delta T}{d}$$

Where:

- $\frac{Q}{t}$= rate of heat transfer (W)

- k= thermal conductivity (W/m·K)

- A= cross-sectional area (m^2)

- ΔT= temperature difference (K)

- d= thickness of material (m)

Convection

Convection is the transfer of heat by the **movement of fluids** (liquids or gases).
When a fluid is heated, it expands, becomes less dense, and rises — while cooler, denser fluid sinks to take its place. This circulation forms **convection currents**.

Examples:

- Boiling water circulating in a pot

- Atmospheric and oceanic currents that drive weather

- Convection inside stars and planets

Convection efficiently transfers energy in systems where fluid motion is possible, redistributing heat throughout the environment.

Radiation

Radiation is the transfer of energy by **electromagnetic waves** — it does not require a medium.
Even in the vacuum of space, energy from the Sun reaches Earth through radiation.

All objects emit radiant energy proportional to the fourth power of their temperature, as described by the **Stefan–Boltzmann Law**:

$$P = \sigma A T^4$$

Where:

- P= radiant power (W)

- $\sigma = 5.67 \times 10^{-8} W/m^2 \cdot K^4$= Stefan–Boltzmann constant

- A= surface area (m^2)

- T= temperature (K)

This process allows stars to shine, the Earth to cool, and every object to exchange energy with its surroundings through light and heat.

8.4 Specific Heat and Calorimetry

Different substances require different amounts of energy to change temperature. This property is called **specific heat capacity (c)**.

Specific Heat Capacity

The specific heat of a substance is the amount of heat required to raise the temperature of **1 kilogram of the substance by 1 kelvin**.

$$Q = mc\Delta T$$

Where:

- Q= heat added or removed (J)
- m= mass (kg)
- c= specific heat capacity (J/kg·K)
- ΔT= temperature change (K)

Substances with high specific heat (like water) can absorb or release large amounts of energy with little temperature change — making them crucial in regulating Earth's climate and biological life.

Substance Specific Heat (J/kg·K)

Substance	Specific Heat (J/kg·K)
Water	4184
Ice	2100
Aluminum	900
Iron	450
Air	1005

Calorimetry

Calorimetry is the experimental measurement of heat transfer.
A **calorimeter** is an insulated device that measures the heat exchanged between substances.

In an ideal calorimeter:

$$Q_{lost} + Q_{gained} = 0$$

This equation allows us to determine unknown quantities like specific heat or heat of reaction by balancing the energy gained and lost in a closed system.

Example:
If hot metal is placed in cool water, the metal loses heat while the water gains heat — until thermal equilibrium is reached.
By measuring temperature changes, one can determine the specific heat of the metal.

8.5 The Microscopic View of Heat

At the molecular level, temperature and heat arise from particle motion. Molecules vibrate, rotate, and translate — exchanging kinetic energy through collisions.

- **Temperature** measures the average kinetic energy per particle.

- **Thermal energy** is the total internal energy.

- **Heat** is energy in transit between systems.

This microscopic understanding links thermodynamics with atomic theory, showing that the macroscopic flow of heat is the collective behavior of countless microscopic interactions.

8.6 The Continuum of Energy

In thermodynamics, energy reveals its universal character.
From the vibration of atoms to the radiation of stars, the principles of temperature and heat govern both the everyday and the cosmic.

Understanding thermal energy bridges the worlds of mechanics, chemistry, and astrophysics — for wherever energy flows, the laws of thermodynamics guide it.

The study of heat is not merely about warmth; it is about **motion, equilibrium, and the transformation of energy itself** — the heartbeat of the physical universe.

Chapter 9: The Laws of Thermodynamics

Energy is eternal — it can neither be created nor destroyed, only transformed. Yet not all transformations are equal; some are spontaneous, others impossible, and all are guided by a set of profound rules known as the **Laws of Thermodynamics**.

These laws describe how energy flows, changes, and interacts in every physical system. They govern engines and ecosystems, black holes and galaxies, and even the biological processes of life itself. They tell us what can happen, what cannot, and why time seems to move in one direction.

This chapter explores the **Zeroth**, **First**, **Second**, and **Third Laws of Thermodynamics**, along with their implications for energy conservation, entropy, and the fundamental limits of machines.

9.1 The Zeroth Law of Thermodynamics: Defining Temperature

Before heat or energy can be compared, we must understand **temperature** and **thermal equilibrium**.

The **Zeroth Law of Thermodynamics** states:

If two systems are each in thermal equilibrium with a third system, then they are in thermal equilibrium with each other.

This law may seem simple, but it establishes the very concept of **temperature** as a measurable quantity.

Key Ideas:

- **Thermal equilibrium** means no net heat flow between systems.

- Temperature provides a consistent way to compare thermal states.

- It forms the foundation for **thermometers** and the definition of temperature scales (Celsius, Kelvin, Fahrenheit).

Without the Zeroth Law, temperature would have no universal meaning — it is the quiet cornerstone upon which the other laws stand.

9.2 The First Law of Thermodynamics: Conservation of Energy

The **First Law** states that energy is conserved, even when it changes form.

$$\Delta U = Q - W$$

Where:

- ΔU = change in internal energy (J)

- Q = heat added to the system (J)

- W = work done by the system (J)

This law is the **thermal version of the Law of Conservation of Energy**. It tells us that the total energy of a closed system remains constant.

Interpretation:

- If heat flows **into** a system ($Q > 0$), internal energy increases.

- If the system does **work** on its surroundings ($W > 0$), internal energy decreases.

Example:

When gas is heated in a piston:

- Adding heat increases internal energy.

- Some of that energy does work by pushing the piston upward.

Even when energy changes form — from chemical to mechanical, or thermal to electrical — the total remains perfectly conserved.

Energy can flow, but it never disappears.

9.3 The Second Law of Thermodynamics: Entropy and the Arrow of Time

While the First Law tells us energy is conserved, it does not tell us the **direction** in which energy flows.
The **Second Law** does.

In any natural process, the total entropy of an isolated system always increases or remains constant; it never decreases.

Entropy (S)

Entropy is a measure of **disorder, randomness,** or **the number of possible microstates** of a system.
Mathematically, it is defined by Ludwig Boltzmann as:

$$S = k_B \ln \Omega$$

Where:

- S= entropy (in joules per kelvin, J/K)

- k_B= Boltzmann's constant (1.38×10^{-23} J/K)

- Ω= number of microscopic configurations consistent with the system's state

Entropy naturally increases because systems tend to evolve from **ordered** to **disordered**, from **energy concentrated** to **energy dispersed**.

The Direction of Energy Flow

Heat always flows **spontaneously from hot to cold,** never the reverse, unless external work is done.
This one-way flow gives rise to the **arrow of time** — the irreversible progression of events from past to future.

Every real process — from ice melting to stars burning —
increases the universe's entropy.
The Second Law thus gives time a direction and defines
the ultimate fate of energy: it spreads, disperses, and
becomes less available to do work.

Kelvin–Planck Statement

*It is impossible to construct a cyclic process whose sole
result is to convert heat entirely into work.*

No engine can be 100% efficient because some energy is
always lost as waste heat.

9.4 The Third Law of Thermodynamics: The Absolute Zero Limit

The **Third Law** addresses what happens as systems
approach absolute zero temperature.

*As the temperature of a system approaches absolute
zero, the entropy of a perfect crystal approaches zero.*

At **absolute zero (0 K)**:

- Atomic motion nearly ceases.

- The system reaches its minimum possible energy
  state.

- It becomes perfectly ordered, with only one
  possible microstate ($\Omega = 1$).

However, **absolute zero can never be reached** — it is a theoretical limit.
No process can remove the last trace of energy from matter. This law defines the lower boundary of temperature and the unattainable perfection of order.

9.5 Conservation of Energy in Thermal Systems

Thermodynamics merges heat and work into one unified concept of energy.
In every thermal process:

$$Q_{in} = W_{out} + \Delta U$$

This expression shows that energy input as heat becomes:

- **Work**, useful mechanical output, and

- **Internal energy**, stored as molecular motion.

Examples:

- **Steam Engine:** Heat from burning fuel does work by moving pistons.

- **Refrigerator:** Work done by a compressor moves heat from cold to hot.

- **Earth's Climate:** Solar radiation (heat input) drives winds, oceans, and weather.

In all these systems, the First Law holds perfectly — the sum of all energy transfers equals zero when accounting for every form.

9.6 Entropy and the Arrow of Time

Entropy gives time its irreversible flow.
While the equations of mechanics and electromagnetism are time-symmetric, the **Second Law** introduces a preferred direction — forward.

Key Concepts:

- **Low Entropy:** Ordered systems (e.g., ice crystals, young stars).

- **High Entropy:** Disordered systems (e.g., melted water, stellar remnants).

- The universe began in a state of low entropy (high order) and is evolving toward higher entropy (thermal equilibrium).

This increase in entropy defines the **thermodynamic arrow of time** — a profound link between physics and the human experience of past, present, and future.

9.7 Heat Engines and Refrigerators

Heat Engines

A **heat engine** converts thermal energy into mechanical work by exploiting temperature differences.

In every cycle:

1. Heat Q_H is absorbed from a high-temperature reservoir.

2. Part of it is converted into work W.

3. The remainder Q_C is expelled to a low-temperature reservoir.

$$W = Q_H - Q_C$$

The **efficiency** of a heat engine is given by:

$$\eta = \frac{W}{Q_H} = 1 - \frac{Q_C}{Q_H}$$

For an ideal **Carnot engine**, which operates reversibly between temperatures T_H and T_C:

$$\eta_{\text{Carnot}} = 1 - \frac{T_C}{T_H}$$

No real engine can exceed this theoretical efficiency — it defines the upper limit of all machines.

Refrigerators and Heat Pumps

A **refrigerator** operates as a **heat engine in reverse**.
It uses work to transfer heat from a cold body to a warm one, against the natural direction of flow.

$$\text{Coefficient of Performance (COP)} = \frac{Q_C}{W}$$

For a **heat pump** (which warms rather than cools):

$$\text{COP}_{HP} = \frac{Q_H}{W}$$

Both devices rely on the Second Law: without input work, heat cannot spontaneously flow from cold to hot.

9.8 The Profound Unity of the Laws

The four laws of thermodynamics form a complete and harmonious system:

Law	Statement	Principle
Zeroth	Defines temperature and thermal equilibrium	Foundation of temperature measurement
First	Energy is conserved	Conservation of energy
Second	Entropy always increases	Direction of natural processes
Third	Entropy approaches zero at absolute zero	Defines the unattainable limit of order

Together, they explain everything from the behavior of gases to the life cycles of stars and the ultimate fate of the cosmos.

In essence, thermodynamics reveals that while energy is eternal, **its transformations are not without consequence**. Each process leaves behind an increase in entropy — a measure of time's progress and the universe's irreversible unfolding.

9.9 The Cosmic Perspective

The Laws of Thermodynamics reach beyond machines and matter — they define the evolution of the entire universe.
The First Law guarantees that the total energy of the cosmos remains constant.
The Second ensures that this energy disperses, leading to the slow cooling and expansion of the universe — a future sometimes called the **heat death** of the cosmos.

Yet within that gradual decline, stars still form, life still grows, and consciousness still reflects upon the laws themselves — proving that even in entropy's shadow, **order and creativity emerge**.

Thermodynamics is more than the science of heat; it is the study of **change, balance, and destiny** — the grand conversation between energy and time that defines our universe.

Chapter 10: Kinetic Theory of Matter

All matter, no matter how solid or still it appears, is composed of particles in constant motion.

Atoms vibrate, molecules collide, and gases rush and rebound in a ceaseless, invisible dance.

The **Kinetic Theory of Matter** describes this microscopic motion and shows how it gives rise to macroscopic properties such as **pressure, temperature,** and **volume**.

Thermodynamics tells us how heat and energy behave on a large scale. The kinetic theory explains **why** — by linking the laws of heat to the motion of individual particles. In doing so, it unites the atomic world with the visible one, revealing that energy and motion are woven together at every level of reality.

10.1 The Microscopic Model of Gases

The kinetic theory models gases as vast numbers of tiny particles (atoms or molecules) moving freely and colliding elastically with each other and the walls of their container.

Assumptions of the Kinetic Theory

1. A gas consists of a large number of particles in **random motion**.

2. The **volume of the particles** themselves is negligible compared to the space they occupy.

3. There are **no attractive or repulsive forces** between particles (idealized case).

4. Collisions between particles and with the container walls are **perfectly elastic** — no loss of kinetic energy.

5. The **average kinetic energy** of gas particles is proportional to the **absolute temperature (Kelvin)** of the gas.

These assumptions create an idealized yet remarkably accurate picture of gas behavior under many conditions.

Particle Motion and Pressure

When gas particles collide with the walls of their container, they exert tiny impulses of force. The combined effect of trillions of collisions per second produces a measurable **pressure (P)**.

From Newton's laws and kinetic principles:

$$P = \frac{1}{3} \frac{Nm\bar{v^2}}{V}$$

Where:

- P = pressure (Pa)

- N = number of particles

- m = mass of one particle

- $\bar{v^2}$ = mean square speed
- V = volume of gas

This equation connects **microscopic motion** (velocity) with **macroscopic quantities** (pressure and volume). Faster-moving molecules strike more often and with greater force — increasing pressure.

10.2 Pressure, Volume, and Temperature Relationships

The behavior of gases under different conditions of **pressure (P)**, **volume (V)**, and **temperature (T)** has been studied for centuries, leading to several empirical gas laws — all derivable from kinetic theory.

Boyle's Law (1662)

At constant temperature, the **pressure of a gas is inversely proportional to its volume**.

$$P \propto \frac{1}{V} \quad \text{or} \quad PV = \text{constant}$$

As a gas is compressed (smaller V), its particles strike the container walls more frequently, increasing P.

Charles's Law (1787)

At constant pressure, the **volume of a gas is directly proportional to its absolute temperature**.

$$V \propto T \operatorname{or} \frac{V}{T} = \text{constant}$$

Heating a gas increases the average speed of its molecules, causing expansion.

Gay-Lussac's Law (1809)

At constant volume, the **pressure of a gas is directly proportional to its temperature**.

$$P \propto T \operatorname{or} \frac{P}{T} = \text{constant}$$

As molecular motion intensifies, particles hit the container walls more forcefully and frequently.

Avogadro's Law (1811)

At constant temperature and pressure, **equal volumes of gases contain equal numbers of molecules**.

$$V \propto n$$

Where n is the number of moles. This principle led to the concept of the **mole** and the definition of **Avogadro's number**:

$$N_A = 6.022 \times 10^{23} \text{ particles/mol}$$

10.3 The Ideal Gas Law and Real Gases

By combining the individual gas laws, we obtain the **Ideal Gas Law** — a unifying equation describing the state of a gas.

$$PV = nRT$$

Where:

- P= pressure (Pa)
- V= volume (m^3)
- n= number of moles
- R= universal gas constant (8.314 J/mol·K)
- T= temperature (K)

This elegant equation links microscopic motion to macroscopic measurement and applies to most gases at ordinary conditions.

Microscopic Form of the Ideal Gas Law

Using Boltzmann's constant k_B and the number of molecules N:

$$PV = Nk_BT$$

This expresses the connection between individual molecular energy and the gas's overall pressure and volume.

Limitations and Real Gases

The Ideal Gas Law assumes:

- Zero intermolecular forces
- Negligible molecular volume
- Perfectly elastic collisions

However, **real gases** deviate from this behavior at **high pressures** and **low temperatures**, where particles are close enough for intermolecular forces and finite sizes to matter.

The **Van der Waals equation** modifies the ideal gas law to account for these effects:

$$(P + \frac{an^2}{V^2})(V - nb) = nRT$$

Where:

- a corrects for intermolecular attraction
- b corrects for finite molecular volume

This refined model bridges the gap between idealized and real behavior — showing how gases approach condensation into liquids when cooled or compressed.

10.4 Statistical Interpretation of Temperature

Temperature, at its deepest level, is not just a measure of heat — it is a **statistical property of molecular motion**.

In kinetic theory, the **average translational kinetic energy** of a gas molecule is directly proportional to its absolute temperature:

$$\bar{E}_k = \frac{3}{2} k_B T$$

Where:

- $\bar{E}_k$= average kinetic energy per molecule (J)

- $k_B = 1.38 \times 10^{-23}$ J/K= Boltzmann's constant

- T= temperature (K)

This means that **temperature is a measure of the average microscopic energy of motion**.
When you heat a gas, you do not "add heat" in an abstract sense — you increase the random speed of its molecules.

Distribution of Molecular Speeds

Not all molecules move at the same speed. Their velocities follow the **Maxwell–Boltzmann distribution**, which describes the probability of molecules having a particular speed at a given temperature.

Key observations:

- At higher T, the curve flattens and shifts right — more molecules move faster.

- At lower T, the curve narrows and shifts left — molecules move more slowly.

- Some always move very fast, and some very slow — a hallmark of statistical behavior.

Root-Mean-Square (RMS) Speed

The **root-mean-square speed** (v_{rms}) represents the effective average speed of molecules in a gas:

$$v_{rms} = \sqrt{\frac{3RT}{M}}$$

Where:

- R = universal gas constant (J/mol·K)

- T = absolute temperature (K)

- M = molar mass of the gas (kg/mol)

As temperature rises, the RMS speed increases — showing that molecular motion and temperature are inseparable.

10.5 The Bridge Between Micro and Macro

The Kinetic Theory of Matter reveals a profound truth:
The grand laws of thermodynamics — energy, temperature, pressure — are not abstractions, but the collective result of **billions of microscopic motions**.

The collisions of atoms become heat; their speed becomes temperature; their impacts become pressure. Every thermodynamic process, from a boiling kettle to a star's fusion core, is rooted in the same microscopic principles.

10.6 The Unity of Motion and Energy

From Newton's mechanics to Einstein's relativity, motion defines physics.
The Kinetic Theory of Matter shows that even at the smallest scales, **motion is the essence of energy**.

Heat, temperature, and pressure are not separate phenomena but manifestations of molecular motion — the restless dance that sustains the universe's vitality.
In that motion, invisible yet eternal, lies the heartbeat of matter itself.

Part IV – Electromagnetism: The Unity of Electric and Magnetic Forces

Chapter 11: Electric Forces and Fields

In the vast structure of nature, electricity is one of the most fundamental phenomena. Every atom, every spark, and every flash of lightning arises from the mysterious power of **electric charge**.

Electromagnetism connects two seemingly different forces — electricity and magnetism — into a single, unified field that governs everything from the flow of electrons in a wire to the light of the stars. Understanding electricity begins with **charge** and the **forces** it creates.

This chapter introduces the concept of electric charge, the laws governing electric forces, and the invisible fields that shape them. It also explores the nature of conductors, insulators, and devices that store electrical energy — **capacitors** — setting the stage for the unity of electromagnetism to come.

11.1 Electric Charge and Coulomb's Law

The Nature of Electric Charge

Electric charge is a **fundamental property of matter**, much like mass. It comes in two types:

- **Positive charge** (protons)

- **Negative charge** (electrons)

Objects become charged when they gain or lose electrons:

- **Losing electrons** → positively charged

- **Gaining electrons** → negatively charged

The total charge in the universe is always **conserved** — it can neither be created nor destroyed, only transferred.

Charge is **quantized**, meaning it exists in discrete multiples of the elementary charge:

$$e = 1.602 \times 10^{-19} \text{ C}$$

where 1 **coulomb (C)** equals the charge of approximately 6.24×10^{18} electrons.

Coulomb's Law

Charles-Augustin de Coulomb (1785) discovered that the force between two point charges is similar in form to Newton's law of gravitation.

$$F = k_e \frac{|q_1 q_2|}{r^2}$$

Where:

- F = electric force (N)

- q_1, q_2 = magnitudes of the charges (C)

- r = distance between charges (m)

- $k_e = 8.99 \times 10^9 \, \text{N} \cdot \text{m}^2/\text{C}^2 =$ Coulomb's constant

Key Properties:

- The force acts **along the line** joining the charges.

- **Like charges repel, opposite charges attract.**

- The force decreases with the **square of the distance** — an **inverse-square law**, just like gravity.

Comparison: Electric vs. Gravitational Force

Property	Electric Force	Gravitational Force
Acts on	Electric charge	Mass
Can be	Attractive or repulsive	Always attractive
Strength	Very strong	Extremely weak
Equation	$F = k_e \dfrac{q_1 q_2}{r^2}$	$F = G \dfrac{m_1 m_2}{r^2}$

The electric force dominates atomic and molecular interactions, shaping the structure of matter, while gravity governs the motion of celestial bodies. Both follow the same mathematical symmetry — a testament to nature's unity.

11.2 Electric Field and Potential

Charges do not act on each other through direct contact. Instead, they create **electric fields** — invisible regions of influence that permeate space.

The electric field describes the **force per unit charge** at any point around a source charge.

$$E = \frac{F}{q}$$

Where:

- **E**= electric field (N/C)

- **F**= force experienced by a test charge (N)

- q= magnitude of the test charge (C)

For a point charge:

$$E = k_e \frac{|Q|}{r^2}$$

Direction and Representation

- The field points **away** from positive charges and **toward** negative charges.

- Field lines indicate direction and strength:

  - Denser lines = stronger field.

  - Lines never cross.

The concept of the field, first introduced by Michael Faraday, transformed physics — replacing the notion of "action at a distance" with the continuous structure of space filled by invisible forces.

Electric Potential Energy

A charge in an electric field has **potential energy** due to its position. Moving a charge within the field requires work.

The **electric potential energy (U)** between two point charges is:

$$U = k_e \frac{q_1 q_2}{r}$$

Just as gravitational potential energy depends on height, electric potential energy depends on separation in the field.

Electric Potential (V)

The **electric potential** (or **voltage**) at a point is the potential energy per unit charge:

$$V = \frac{U}{q}$$

For a point charge:

$$V = k_e \frac{Q}{r}$$

- Unit: **volt (V)**, where $1\,V = 1\,J/C$.

- Electric potential is a **scalar quantity**, unlike the vector field **E**.

The relationship between electric field and potential is:

$$E = -\nabla V$$

meaning the electric field points in the direction of **decreasing potential.**

Equipotential Surfaces

Equipotential surfaces connect points with the same electric potential.

- No work is done moving a charge along an equipotential surface.

- Electric field lines always cross equipotential surfaces at right angles.

These surfaces visualize the energy landscape of electric fields — from the symmetric fields around point charges to the complex fields near conductors and capacitors.

11.3 Conductors, Insulators, and Capacitors

The way materials respond to electric charge defines their role in electrical systems. Some allow charge to flow freely, while others resist or store it.

Conductors

Conductors contain **free electrons** that can move easily through the material (typically metals).

Properties:

- Charge resides **on the surface** of a conductor in electrostatic equilibrium.

- The electric field **inside a conductor** is **zero**.

- Conductors can **shield** their interiors from external electric fields (a **Faraday cage**).

Examples: copper, aluminum, silver.

Insulators

Insulators (or **dielectrics**) do not allow charges to move freely.
Their electrons are tightly bound to atoms, making them poor conductors of electricity.

Examples: rubber, glass, plastic, air.

However, when placed in an electric field, insulators become **polarized** — their charges shift slightly, reducing the effective field inside. This behavior is vital for capacitors and electric insulation.

Capacitors

A **capacitor** is a device that **stores electric charge and energy** in an electric field.

It consists of two **conducting plates** separated by an **insulating dielectric**.

When a voltage V is applied across the plates, opposite charges accumulate on each plate, creating an electric field between them.

The **capacitance (C)** measures how much charge a capacitor can store per unit voltage:

$$C = \frac{Q}{V}$$

Where:

- C = capacitance (in farads, F)

- Q = stored charge (C)

- V = potential difference (V)

For a parallel-plate capacitor:

$$C = \varepsilon_0 \frac{A}{d}$$

Where:

- $\varepsilon_0 = 8.85 \times 10^{-12}$ F/m (vacuum permittivity)

- A = plate area (m^2)

- d = separation distance (m)

Energy Stored in a Capacitor

The energy stored in a charged capacitor is given by:

$$U = \frac{1}{2}CV^2$$

This energy resides in the **electric field** between the plates and can be released when the capacitor discharges.

Capacitors play essential roles in circuits — storing energy, smoothing current, and timing electrical pulses. They also model the way nature stores energy in fields, from lightning clouds to the membranes of living cells.

11.4 The Field as a Bridge Between Matter and Energy

The concept of the **electric field** marks one of the most profound shifts in physics. It transforms our understanding of the universe from one of isolated particles to one of continuous interaction — a fabric woven by energy itself.

In every spark, lightning bolt, and electric current, these same principles operate.
The forces of attraction and repulsion between charges shape atoms, chemical bonds, and the flow of energy that powers civilization.

Electromagnetism begins here — with the unseen geometry of electric fields. In the chapters that follow, we will see how moving charges give rise to **magnetism**, and how both phenomena unite to create one of nature's greatest symmetries: **light itself.**

Chapter 12: Electric Circuits

Electricity is the lifeblood of modern civilization. From the glowing filaments of a lamp to the complex circuits in computers, the controlled flow of electric charge enables nearly every technology we use.

An **electric circuit** is the pathway through which charges move under the influence of an electric potential difference. By understanding how current, voltage, and resistance interact, we uncover the principles that govern everything from a simple battery-powered light to the vast power grids of the world.

This chapter introduces the fundamental concepts of electric current, voltage, resistance, and the laws that relate them. It also explores how components combine in series and parallel circuits, and how electrical energy and power are generated, transmitted, and used.

12.1 Current, Voltage, and Resistance

Electric Current (I)

Electric current is the rate at which electric charge flows through a conductor.

$$I = \frac{\Delta Q}{\Delta t}$$

Where:

- I = current (in amperes, A)

- ΔQ= charge passing through a point (in coulombs, C)

- Δt= time interval (in seconds, s)

1 **ampere (A)** equals 1 **coulomb per second (C/s)**.

Current flows from **high potential to low potential**, although the physical flow of electrons is in the opposite direction (from negative to positive). This convention, called **conventional current**, dates back to Benjamin Franklin.

Voltage (V)

Voltage, or **electric potential difference**, is the energy per unit charge that drives electric current through a circuit.

$$V = \frac{W}{Q}$$

Where:

- V= voltage (in volts, V)

- W= work or energy (in joules, J)

- Q= charge (in coulombs, C)

A **1-volt** potential difference means **1 joule** of energy is used to move **1 coulomb** of charge between two points.

Voltage can be thought of as the "pressure" pushing charges through a conductor.

Resistance (R)

All materials oppose the flow of electric current to some degree. This opposition is called **electrical resistance.**

$$R = \frac{V}{I}$$

Where:

- R = resistance (in ohms, Ω)
- V = voltage (V)
- I = current (A)

Resistance depends on four key factors:

1. **Material** — good conductors like copper have low resistance.

2. **Length (L)** — longer wires have greater resistance.

3. **Cross-sectional area (A)** — thicker wires have less resistance.

4. **Temperature** — resistance increases with temperature in most conductors.

$$R = \rho \frac{L}{A}$$

Where ρ is the **resistivity** of the material ($\Omega \cdot$m), a constant characteristic of each substance.

12.2 Ohm's Law

Discovered by Georg Simon Ohm in 1827, **Ohm's Law** defines the linear relationship between voltage, current, and resistance in a conductor:

$$V = IR$$

This simple but powerful equation forms the cornerstone of circuit theory.

Interpretation:

- At constant resistance, increasing voltage increases current.

- For a given voltage, increasing resistance decreases current.

- A plot of V vs. I for an ohmic conductor is a straight line, with the slope equal to R.

Ohmic materials (such as metals) obey Ohm's Law. **Non-ohmic materials** (such as diodes and semiconductors) do not — their resistance changes with voltage or temperature.

Example

A 12 V battery drives 3 A of current through a resistor. The resistance is:

$$R = \frac{V}{I} = \frac{12}{3} = 4\,\Omega$$

If another resistor of 8 Ω is connected to the same battery, the current becomes:

$$I = \frac{V}{R} = \frac{12}{8} = 1.5\text{ A}$$

Thus, greater resistance results in less current flow.

12.3 Series and Parallel Circuits

Electrical components can be connected in two main ways: **series** (one after another) and **parallel** (side by side). Understanding these arrangements is crucial for analyzing real-world circuits.

Series Circuits

In a **series circuit**, current has only one path to follow. All components share the same current, but the total voltage divides among them.

$$I_1 = I_2 = I_3 = \cdots = I$$
$$V_{total} = V_1 + V_2 + V_3 + \cdots$$
$$R_{total} = R_1 + R_2 + R_3 + \cdots$$

Key points:

- A break anywhere in the circuit stops all current.

- Used where constant current is needed (e.g., old-style Christmas lights).

Example:
Three resistors (2 Ω, 3 Ω, 5 Ω) connected in series:

$$R_{total} = 2 + 3 + 5 = 10 \, \Omega$$

Parallel Circuits

In a **parallel circuit**, each component is connected across the same two points, providing multiple current paths.
The voltage across each branch is the same, but current divides according to resistance.

$$V_1 = V_2 = V_3 = \cdots = V_{total}$$
$$\frac{1}{R_{total}} = \frac{1}{R_1} + \frac{1}{R_2} + \frac{1}{R_3} + \cdots$$

Key points:

- If one branch fails, current continues through others.

- Parallel wiring is used in homes so each device gets full voltage.

Example:

Two resistors (4 Ω and 12 Ω) connected in parallel:

$$\frac{1}{R_{total}} = \frac{1}{4} + \frac{1}{12} = \frac{1}{3} \Rightarrow R_{total} = 3 \, \Omega$$

The total resistance of a parallel circuit is **always less** than the smallest individual resistance.

Combination Circuits

Most real circuits combine both series and parallel elements.

They can be simplified step by step using the rules above — adding series resistances and combining parallel ones — until a single equivalent resistance is found.

This process allows calculation of current, voltage, and power throughout the circuit.

12.4 Electric Power and Energy

Electric Power

Power (P) is the rate at which electrical energy is converted into another form — mechanical work, light, or heat.

$$P = IV$$

Using Ohm's Law, power can also be expressed as:

$$P = I^2 R = \frac{V^2}{R}$$

Where:

- P = power (in watts, W)
- I = current (A)
- V = voltage (V)
- R = resistance (Ω)

One **watt (W)** equals one **joule per second (J/s)**.

Examples of power conversion:

- Electric heaters: electrical → thermal energy
- Light bulbs: electrical → light + heat
- Motors: electrical → mechanical energy

Electrical Energy

Energy consumption in electrical systems depends on both power and time.

$$E = Pt$$

Where:

- E = energy (in joules, J)
- P = power (W)

- t = time (s)

In everyday life, electrical energy is measured in **kilowatt-hours (kWh)**:

$$1 \text{ kWh} = 3.6 \times 10^6 \text{ J}$$

For example, a 100 W light bulb operating for 10 hours uses:

$$E = 0.1 \text{ kW} \times 10 \text{ h} = 1 \text{ kWh}$$

Efficiency and Energy Loss

No device converts energy perfectly. Some electrical energy is always lost as heat due to resistance.

$$\text{Efficiency} = \frac{\text{Useful Power Output}}{\text{Total Power Input}} \times 100\%$$

Reducing resistance, using better materials, and managing heat are central to improving electrical efficiency — from household appliances to high-voltage transmission lines.

12.5 The Flow of Energy Through Circuits

Electric circuits embody the flow of energy itself.
A battery provides **electric potential**, charges carry that

energy through conductors, and resistors or devices **transform** it into light, heat, or motion.

Like rivers of charge, circuits illustrate how energy moves in closed loops — continuously converted, but never lost.

From the lightning in a storm to the hum of a generator, all electrical phenomena obey the same elegant principles: **Voltage drives. Resistance resists. Current flows. Energy transforms.**

Chapter 13: Magnetism and Electromagnetic Induction

In nature, electricity and magnetism are two faces of the same force. A moving charge produces a magnetic field, and a changing magnetic field produces an electric current. This dynamic interplay — **electromagnetic induction** — powers the world's generators, motors, and transformers, and forms the very essence of light itself.

This chapter explores the origin of magnetic fields, the forces they exert, the motion of charged particles in those fields, and the great discoveries of **Faraday** and **Lenz**, whose laws describe how changing fields give rise to electric currents.

13.1 Magnetic Fields and Forces

The Nature of Magnetism

Magnetism arises from the motion of electric charges. On the atomic scale, magnetic fields originate from the **spins** and **orbital motion** of electrons around nuclei.
Every magnet has two poles — **north (N)** and **south (S)** — which always appear in pairs. Breaking a magnet in half produces two smaller magnets, each with its own poles.

Magnetic Field (B)

The **magnetic field** describes the influence that a magnetic source exerts on other magnets or moving charges in the surrounding space.

$$F = q\mathbf{v} \times \mathbf{B}$$

Where:

- **F**= magnetic force on a charge (N)

- q= charge (C)

- **v**= velocity of the charge (m/s)

- **B**= magnetic field (T, tesla)

The direction of **F** is given by the **right-hand rule**: Point your fingers in the direction of velocity (**v**), curl them toward the magnetic field (**B**), and your thumb points in the direction of the force on a **positive charge**. (It's reversed for negative charges.)

Magnetic Field Lines

Magnetic fields can be visualized as lines that:

- Emerge from the **north pole** and enter the **south pole**.

- Form closed loops — even inside the magnet itself.

- Never intersect.

- Indicate strength by their density (closer lines mean stronger fields).

Earth itself is a giant magnet, with a magnetic field extending from pole to pole, shielding life from solar radiation and guiding compasses for centuries.

Magnetic Force on a Current-Carrying Wire

A current in a wire experiences a magnetic force because moving charges interact with magnetic fields.

$$\mathbf{F} = I\mathbf{L} \times \mathbf{B}$$

Where:

- I= current (A)

- $\mathbf{L}$= length vector of the wire (m)

- $\mathbf{B}$= magnetic field (T)

This principle underlies **electric motors**, where magnetic forces produce motion.

13.2 Motion of Charged Particles in Fields

When a charged particle moves through a magnetic field, it experiences a force perpendicular to both its velocity and the field — causing **circular or helical motion**.

Centripetal Motion

For a particle of charge q, mass m, and velocity v, moving perpendicular to a magnetic field B:

$$F = qvB = \frac{mv^2}{r}$$

Solving for the radius:

$$r = \frac{mv}{qB}$$

- Particles with greater momentum or smaller charge curve less (larger r).

- This principle is used in **mass spectrometers** and **cyclotrons** to separate charged particles by mass and charge.

If the velocity is not perpendicular, the motion becomes **helical** — a combination of circular motion and forward drift along the magnetic field lines.

Velocity Selector

By combining **electric** and **magnetic** fields, one can select particles of a specific velocity.
In such a device:

$$qE = qvB \Rightarrow v = \frac{E}{B}$$

This principle is used in **cathode-ray tubes** and **particle accelerators**.

13.3 Faraday's Law of Electromagnetic Induction

Faraday's Great Discovery

In 1831, **Michael Faraday** discovered that **a changing magnetic field induces an electric current** in a nearby conductor.
This phenomenon, called **electromagnetic induction**, demonstrated that electricity and magnetism are fundamentally linked.

Faraday's Law

$$\varepsilon = -\frac{d\Phi_B}{dt}$$

Where:

- ε = induced electromotive force (emf, in volts)

- Φ_B = magnetic flux through the circuit (Wb, weber)

Magnetic flux is defined as:

$$\Phi_B = BA\cos(\theta)$$

Where:

- B = magnetic field (T)

- A = area of the loop (m^2)

- θ = angle between **B** and the normal to the surface

Faraday's Law shows that an emf is produced whenever the magnetic flux changes — by altering B, A, or θ.

It is this principle that makes **generators**, **transformers**, and **induction coils** possible.

Lenz's Law

Heinrich Lenz refined Faraday's discovery by determining the **direction** of the induced current.

The direction of the induced current is such that it opposes the change in magnetic flux that produced it.

This opposition is expressed by the **negative sign** in Faraday's equation.
It reflects **energy conservation**: the induced current always acts to resist the change that caused it.

Example:
If a magnetic field through a loop increases, the induced current creates its own magnetic field in the **opposite direction** to oppose the increase.

13.4 Generators, Motors, and Transformers

The practical power of electromagnetism comes from its ability to convert between **mechanical** and **electrical** energy.

Electric Generators

A **generator** converts mechanical energy into electrical energy through electromagnetic induction.

When a coil of wire rotates in a magnetic field, the changing flux induces an alternating voltage:

$$\varepsilon = \varepsilon_{max}\sin(\omega t)$$

Where:

- ω = angular velocity of the coil

- ε_{max} = maximum emf

Generators are the heart of modern power systems — turning the rotation of turbines driven by wind, water, or steam into the electricity that lights the world.

Electric Motors

A **motor** performs the reverse process: it converts electrical energy into mechanical motion.

When current flows through a coil in a magnetic field, a torque is produced:

$$\tau = NBIA\sin(\theta)$$

Where:

- N = number of turns

- B = magnetic field

- I = current

- A = area of the coil

- θ = angle between coil and field

This torque causes the coil to rotate — transforming electric energy into motion.
From fans and vehicles to robotics, motors translate the invisible flow of charge into physical work.

Transformers

A **transformer** transfers electrical energy between two circuits by electromagnetic induction. It consists of two coils — a **primary** and a **secondary** — wound around a common magnetic core.

$$\frac{V_s}{V_p} = \frac{N_s}{N_p}$$

Where:

- V_p, V_s = primary and secondary voltages

- N_p, N_s = number of turns in the coils

A **step-up transformer** increases voltage; a **step-down transformer** decreases it.
Energy is conserved, so if voltage increases, current decreases proportionally:

$$V_p I_p = V_s I_s$$

Transformers allow the efficient transmission of electricity over long distances — a cornerstone of modern power grids.

13.5 The Unity of Electricity and Magnetism

Faraday and Lenz revealed that motion and change intertwine — a moving charge creates a magnetic field, and a changing magnetic field creates an electric field.

Decades later, **James Clerk Maxwell** united these phenomena in a single mathematical framework: **Maxwell's Equations**, which showed that electric and magnetic fields are not separate, but part of a single entity — the **electromagnetic field**.

From this union arises one of nature's greatest revelations:
Light itself is an electromagnetic wave — oscillating electric and magnetic fields traveling through space at the speed of $c = 3.00 \times 10^8$ m/s.

13.6 The Power of Induction

Electromagnetism defines modern civilization — from the current in our homes to the signals in our communication networks. Every generator, motor, and transformer is a direct application of the laws discovered by Faraday, Lenz, and Maxwell.

Through them, humanity learned to harness one of the most fundamental forces of the universe — to light our cities, move our machines, and send our voices across the stars.

Magnetism and induction remind us that **motion and change** are the heart of physical reality — that energy flows, fields interact, and the universe itself is alive with movement.

Chapter 14: Maxwell's Equations and Electromagnetic Waves

In the mid-19th century, physics witnessed a revolution. Through the work of **James Clerk Maxwell**, the separate forces of **electricity** and **magnetism** were revealed to be two aspects of the same underlying phenomenon — the **electromagnetic field**.

Maxwell's insights unified Faraday's lines of force, Ampère's currents, and Gauss's flux laws into four mathematical statements of extraordinary power and beauty. From these equations emerged a startling conclusion: **light itself is an electromagnetic wave**, traveling through space at a finite speed.

This chapter explores Maxwell's unification, the nature of electromagnetic waves, the electromagnetic spectrum, and how light interacts with matter through reflection, refraction, and interference — phenomena that define both vision and technology.

14.1 Unifying Electricity and Magnetism

Before Maxwell, electricity and magnetism were studied as separate forces:

- **Coulomb's Law** described electric forces between charges.

- **Faraday's Law** described how changing magnetic fields induce electric currents.

- **Ampère's Law** described magnetic fields created by electric currents.

Maxwell recognized that these were parts of a **single system**. When combined and slightly modified, they formed the foundation of **classical electromagnetism**.

Maxwell's Four Equations (in integral form)

1. **Gauss's Law for Electricity**

$$\oint \mathbf{E} \cdot d\mathbf{A} = \frac{Q_{enclosed}}{\varepsilon_0}$$

Electric charges produce electric fields.
The total electric flux through a closed surface equals the enclosed charge divided by the permittivity of free space (ε_0).

2. **Gauss's Law for Magnetism**

$$\oint \mathbf{B} \cdot d\mathbf{A} = 0$$

There are no magnetic monopoles — magnetic field lines form continuous loops; every north pole is paired with a south pole.

3. **Faraday's Law of Induction**

$$\oint \mathbf{E} \cdot d\mathbf{s} = -\frac{d\Phi_B}{dt}$$

A changing magnetic field induces an electric field. This is the principle behind generators and transformers.

4. Ampère–Maxwell Law

$$\oint \mathbf{B} \cdot d\mathbf{s} = \mu_0 I + \mu_0 \varepsilon_0 \frac{d\Phi_E}{dt}$$

Magnetic fields arise from both electric currents (I) and changing electric fields ($\frac{d\Phi_E}{dt}$).

Maxwell's addition of the **displacement current term** ($\mu_0 \varepsilon_0 \frac{d\Phi_E}{dt}$) completed the symmetry between electric and magnetic phenomena.

The Profound Consequence

From these four equations, Maxwell derived a single, astonishing result: **a changing electric field generates a magnetic field, and a changing magnetic field generates an electric field.**

These two effects sustain each other, allowing a disturbance — an **electromagnetic wave** — to travel through space even in the absence of matter.

By combining the equations, Maxwell found that such waves propagate at a speed:

$$c = \frac{1}{\sqrt{\mu_0 \varepsilon_0}} \approx 3.00 \times 10^8 \text{ m/s}$$

This value matched the known speed of light, leading Maxwell to conclude that **light is an electromagnetic phenomenon**.

14.2 Light as an Electromagnetic Wave

Light consists of oscillating **electric (E)** and **magnetic (B)** fields, perpendicular to each other and to the direction of wave propagation.

$$\mathbf{E} \perp \mathbf{B} \perp \mathbf{k}$$

These oscillations move together through space, carrying **energy**, **momentum**, and **information**. The electric field drives charges, while the magnetic field arises from their motion — each creating and sustaining the other.

Wave Equation

The electromagnetic wave satisfies:

$$\nabla^2 \mathbf{E} = \mu_0 \varepsilon_0 \frac{\partial^2 \mathbf{E}}{\partial t^2}$$

and similarly for **B**.

The solution represents a wave traveling at velocity c, with wavelength λ, frequency f, and energy:

$$E = hf$$

where $h = 6.626 \times 10^{-34}$ J · s Planck's constant.

Energy and Intensity

The **energy density** of an electromagnetic wave is shared equally between electric and magnetic fields:

$$u = \frac{1}{2}\varepsilon_0 E^2 + \frac{1}{2}\frac{B^2}{\mu_0}$$

The **intensity (I)**, or power per unit area, is:

$$I = \langle S \rangle = \frac{1}{2}c\varepsilon_0 E_0^2$$

where $S = E \times H$ is the **Poynting vector**, representing the direction of energy flow.

14.3 The Electromagnetic Spectrum

Electromagnetic waves encompass a vast range of frequencies and wavelengths, far beyond the narrow band visible to the human eye.

Region	Wavelength Range	Frequency Range	Common Sources / Applications
Radio waves	> 1 m	< 300 MHz	Broadcasting, communication

Region	Wavelength Range	Frequency Range	Common Sources / Applications
Microwaves	1 mm – 1 m	300 MHz – 300 GHz	Radar, ovens, wireless signals
Infrared (IR)	700 nm – 1 mm	300 GHz – 400 THz	Heat radiation, remote controls
Visible light	400 – 700 nm	400 – 750 THz	Human vision, optical devices
Ultraviolet (UV)	10 – 400 nm	750 THz – 30 PHz	Sterilization, fluorescence
X-rays	0.01 – 10 nm	30 PHz – 30 EHz	Medical imaging, crystallography
Gamma rays	< 0.01 nm	> 30 EHz	Nuclear decay, cosmic phenomena

All electromagnetic waves travel at the same speed in a vacuum — $c = 3.00 \times 10^8$ m/s— but differ in energy and interaction with matter.

From the faint cosmic background to the radiation of stars, this spectrum is the universal language of light.

14.4 Reflection, Refraction, and Interference

When light encounters matter, it interacts through processes that reveal its wave nature and the laws of optics.

Reflection

When light strikes a surface, part of it **reflects** according to the **law of reflection**:

$$\theta_i = \theta_r$$

where θ_i is the angle of incidence and θ_r the angle of reflection, both measured relative to the normal.

Reflection can be:

- **Specular** (smooth surfaces, like mirrors)
- **Diffuse** (rough surfaces, scattering light in many directions)

Reflection underlies mirrors, optical instruments, and even the twinkling of stars on rippling water.

Refraction

When light passes between media of different optical densities, its speed changes, bending the path — a phenomenon called **refraction**.

Described by **Snell's Law**:

$$n_1 \sin \theta_1 = n_2 \sin \theta_2$$

Where:

- $n = \frac{c}{v}$ is the **index of refraction** of the medium.

- θ_1, θ_2 = angles of incidence and refraction.

Applications:

- Lenses in cameras, telescopes, and eyes

- Rainbows (dispersion of light through water droplets)

- Fiber-optic communication, where light is guided by **total internal reflection**

Interference

Interference occurs when two or more waves overlap, producing regions of **constructive** (bright) and **destructive** (dark) superposition.

Conditions for interference depend on **coherence** — the stability of phase between waves.

Examples:

- **Double-slit experiment:** Bright and dark fringes reveal light's wave nature.

- **Thin-film interference:** Colorful patterns in soap bubbles and oil films result from constructive and destructive interference of reflected light waves.

Interference demonstrates the principle that light, though composed of photons, behaves as a **wave** when propagating — one of the great dualities of nature.

14.5 The Unity of Light and Fields

Maxwell's equations not only united electricity and magnetism — they also revealed that **light, radio waves, and all forms of radiation** are manifestations of a single entity: the **electromagnetic field**.

This discovery transformed physics and laid the foundation for quantum mechanics, relativity, and modern technology — from lasers and communication to particle accelerators and cosmology.

In Maxwell's elegant synthesis, we see nature's harmony: Electricity creates magnetism. Magnetism creates electricity.
Together, they create **light** — the bridge between energy and matter, between the human eye and the cosmos itself.

Part V – Relativity: The Geometry of Space and Time

Chapter 15: Special Relativity

At the dawn of the 20th century, physics stood at a crossroads. Newton's laws had reigned supreme for over two centuries, describing motion and gravity with unmatched precision. Yet a silent conflict had emerged between **Newtonian mechanics** and **Maxwell's equations** of electromagnetism.

The speed of light, $c = 3.00 \times 10^8$ m/s, appeared constant in all experiments — regardless of the motion of the source or observer. This contradicted the classical idea that velocities simply add.

In 1905, **Albert Einstein**, then a young patent clerk in Bern, proposed a bold resolution. He reimagined the nature of space and time themselves, merging them into a single fabric — **spacetime** — where the laws of physics are the same for all observers moving uniformly relative to one another.

His theory, **Special Relativity**, forever changed our understanding of motion, time, energy, and the very structure of the universe.

15.1 The Postulates of Relativity

Einstein built his theory on two simple but revolutionary postulates.

1. The Principle of Relativity

The laws of physics are the same in all inertial (non-accelerating) reference frames.

No experiment conducted in uniform motion can determine whether you are truly "at rest" or "in motion." Motion is **relative** — only the relative velocity between observers has physical meaning.

This principle extends Galileo's relativity (which applied to mechanics) to **all** of physics, including electromagnetism.

---

2. The Constancy of the Speed of Light

The speed of light in a vacuum is the same for all observers, regardless of their motion or the motion of the light source.

No matter how fast you move toward or away from a beam of light, you will always measure its speed as c. This postulate shattered classical intuition, requiring a complete rethinking of time, space, and simultaneity.

Consequences

From these two postulates flow all the counterintuitive results of relativity:

- Time slows down for moving clocks.

- Lengths contract in the direction of motion.

- Mass and energy are equivalent.

- Simultaneity is no longer absolute — what is "now" for one observer may not be "now" for another.

Einstein's insight revealed that **space and time are interwoven**, forming a four-dimensional continuum governed by the geometry of motion.

15.2 Time Dilation and Length Contraction

Time Dilation

Moving clocks run more slowly when observed from a stationary frame.

If a clock moves at speed v relative to an observer, the time interval Δt measured by the observer is longer than the time Δt_0 measured by the moving clock:

$$\Delta t = \frac{\Delta t_0}{\sqrt{1 - \frac{v^2}{c^2}}}$$

Where:

- Δt_0 = proper time (measured in the clock's rest frame)

- Δt = dilated time (measured by the observer)

As v approaches c, time dilation becomes significant. At $v = 0.99c$, moving clocks tick at about one-seventh their normal rate.

Example:
Muons created in Earth's atmosphere travel toward the surface. Their short lifetime (microseconds) should make it impossible for them to reach the ground — yet they do. From the Earth's frame, their internal "clocks" run slow due to relativistic time dilation.

Length Contraction

Objects moving at high speeds appear shorter **along the direction of motion** to a stationary observer.

$$L = L_0 \sqrt{1 - \frac{v^2}{c^2}}$$

Where:

- L_0 = proper length (measured in the object's rest frame)

- L = contracted length (measured by the observer)

Length contraction and time dilation are two sides of the same phenomenon — the geometry of spacetime. Neither effect is due to mechanical compression or slowing; both arise naturally from the structure of the universe itself.

The Lorentz Factor

Both time dilation and length contraction depend on the **Lorentz factor:**

$$\gamma = \frac{1}{\sqrt{1 - \frac{v^2}{c^2}}}$$

As $v \rightarrow c, \gamma \rightarrow \infty$.

No object with mass can reach the speed of light, because it would require infinite energy.

15.3 Relativistic Momentum and Energy

At relativistic speeds, Newton's definitions of momentum and energy must be modified.

The **relativistic momentum** is given by:

$$p = \gamma m v$$

Where:

- m = rest mass
- v = velocity
- γ = Lorentz factor

Momentum increases dramatically as v approaches c, ensuring that massive objects can never attain light speed.

Total Energy

Einstein found that the total energy of a particle includes both rest and motion components:

$$E = \gamma mc^2$$

When the particle is at rest ($v = 0$):

$$E_0 = mc^2$$

This is the **rest energy**, the energy an object possesses by virtue of its mass alone. It implies that **mass is a concentrated form of energy** — a concept that changed both science and civilization.

Kinetic Energy

The relativistic kinetic energy is:

$$K = (\gamma - 1)mc^2$$

At low speeds ($v \ll c$), this reduces to the classical form:

$$K \approx \frac{1}{2}mv^2$$

showing that classical mechanics is a special case of relativity — an approximation valid at everyday speeds.

15.4 $E = mc^2$: Mass–Energy Equivalence

The most famous equation in science,

$$E = mc^2$$

states that **mass and energy are interchangeable**.

A small amount of mass can be converted into a tremendous amount of energy because the conversion factor, c^2, is enormous (9×10^{16} m^2/s^2).

Implications

- **Nuclear Reactions:**
  In fission and fusion, small differences in mass between reactants and products are released as energy.
  This principle powers the Sun and all stars — as well as nuclear reactors and weapons.

- **Particle Physics:**
  High-energy collisions in accelerators convert kinetic energy into new particles, demonstrating that energy can *create* mass.

- **Cosmology:**
  The early universe transformed pure radiation into matter, showing that energy and mass are two forms of the same cosmic currency.

The Unity of Matter and Energy

Einstein's discovery erased the boundary between what things *are* and what they *do*. Matter is energy condensed into stable form; energy is matter in motion.

This equivalence reshaped humanity's understanding of existence — from the fires of stars to the structure of atoms — and became the cornerstone of modern physics.

15.5 The Geometry of Spacetime

Special Relativity unites space and time into a four-dimensional continuum — **spacetime**.
Every event has coordinates (x, y, z, t), and the "distance" between events is described not by Euclidean geometry, but by the **spacetime interval**:

$$s^2 = c^2 t^2 - x^2 - y^2 - z^2$$

This quantity is invariant — it remains the same for all observers, regardless of their relative motion.
Spacetime replaces the rigid stage of Newton's universe with a dynamic geometry that bends and stretches with motion and light.

This insight paved the way for Einstein's later masterpiece — the **General Theory of Relativity**, where gravity itself is understood as the curvature of spacetime.

15.6 The Legacy of Special Relativity

Special Relativity redefined physics, philosophy, and the human sense of reality.
It showed that:

- Time and space are relative, not absolute.

- Energy and mass are two sides of one coin.

- Light defines the structure and limit of causality.

From GPS satellites to particle accelerators, its principles shape the technologies of the modern world.
More profoundly, it revealed that **the laws of the universe are woven into the geometry of spacetime** — elegant, consistent, and universal.

Einstein's vision transformed not only science but our understanding of existence itself — showing that the universe is not a clockwork machine, but a living geometry of motion, energy, and light.

Chapter 16: General Relativity and Gravitation

When Einstein published his **Special Theory of Relativity** in 1905, it transformed our understanding of motion, time, and energy. But one profound force remained beyond its reach — **gravity**.

For centuries, gravity was seen as an invisible force acting at a distance, as described by **Newton's Law of Universal Gravitation**. Though astonishingly successful, Newton's theory could not explain why gravity acts instantly across space, nor could it fit within the constant speed of light demanded by relativity.

After a decade of deep reflection, Einstein unveiled his **General Theory of Relativity** in 1915 — a breathtaking unification of gravity, geometry, and spacetime itself. In this theory, gravity is not a force pulling on objects, but the **curvature of spacetime** caused by mass and energy.

Space and time are no longer the stage on which events unfold — they are active participants in the cosmic drama.

16.1 Gravity as the Curvature of Spacetime

In Einstein's vision, **mass and energy tell spacetime how to curve**, and **spacetime tells matter how to move**.

A massive object like the Sun warps the geometry of spacetime around it. Planets follow the straightest possible paths (called **geodesics**) through this curved geometry — what we perceive as **orbiting under the influence of gravity**.

Mathematically, this relationship is expressed in the **Einstein Field Equations**:

$$G_{\mu\nu} = \frac{8\pi G}{c^4} T_{\mu\nu}$$

Where:

- $G_{\mu\nu}$ describes the curvature of spacetime (geometry).

- $T_{\mu\nu}$ describes the energy and momentum of matter (content).

- G = gravitational constant.

- c = speed of light.

This equation encapsulates the profound unity between geometry and physics — between the shape of space and the flow of energy.

Visualizing Curvature

Imagine placing a heavy ball on a stretched rubber sheet. The sheet curves around the mass, and smaller balls rolling nearby spiral inward, following the curve.

Similarly, the Sun curves spacetime around it, and Earth moves along a geodesic in that curved geometry. There is no mysterious force pulling the Earth — it simply follows the "straightest" path possible through spacetime's distortion.

Gravitational Time Dilation

Because time and space are linked, gravity affects not just motion but the **flow of time** itself.
Clocks run slower in stronger gravitational fields — a phenomenon confirmed by experiments on Earth and near massive celestial bodies.

$$t_{observer} = t_0 \sqrt{1 - \frac{2GM}{rc^2}}$$

This means time passes more slowly near a massive object. At Earth's surface, the effect is small but measurable; near a black hole, it becomes dramatic — time can virtually stand still.

16.2 The Equivalence Principle

Einstein's path to General Relativity began with a simple thought experiment.

The Principle

No local experiment can distinguish between uniform acceleration and a uniform gravitational field.

If you are inside a sealed elevator in deep space, accelerating upward at 9.8 m/s^2, you will feel a "force" pressing you to the floor — indistinguishable from standing on Earth's surface under gravity.
Likewise, in free fall, you feel weightless — even though gravity acts upon you.

This **equivalence principle** suggests that gravity and acceleration are two aspects of the same reality.
It implies that gravity is not a true force, but a manifestation of curved spacetime — where freely falling objects move along natural geodesics determined by geometry, not by an external pull.

Light and Gravity

If gravity affects time, it must also affect light.
Einstein predicted that light passing near a massive body would **bend** due to spacetime curvature.

This was confirmed in 1919, when **Sir Arthur Eddington** observed starlight deflected around the Sun during a solar eclipse. The measurements matched Einstein's predictions perfectly, catapulting him to international fame.

Light's bending by gravity not only proved General Relativity but also revealed that **space and time are inseparable aspects of one geometric fabric**.

16.3 Black Holes and Gravitational Waves

Einstein's equations predicted extraordinary phenomena far beyond his own imagination — **black holes** and **gravitational waves** — both later confirmed by observation.

Black Holes

A **black hole** is a region of spacetime where gravity is so strong that nothing, not even light, can escape.
When a massive star exhausts its nuclear fuel, it collapses under its own gravity. If the mass exceeds a certain limit, no known force can halt the collapse, forming a singularity surrounded by an **event horizon** — the boundary beyond which escape is impossible.

The radius of this event horizon is given by the **Schwarzschild radius**:

$$r_s = \frac{2GM}{c^2}$$

At this boundary, spacetime curves infinitely, and the known laws of physics break down.
Black holes can merge, spin, and devour matter, producing immense energy and shaping galaxies across cosmic time.

Modern astronomy detects black holes through their gravitational effects and by observing X-rays and radio waves emitted as matter spirals inward — revealing these invisible giants at the hearts of galaxies.

Gravitational Waves

Einstein also predicted that accelerating masses would produce **ripples in spacetime**, traveling outward at the speed of light — **gravitational waves**.

In 2015, a century after his prediction, the **LIGO** observatory detected these waves from two merging black holes. The waves carried energy across billions of light-years, briefly stretching and squeezing space itself as they passed through Earth.

Gravitational waves confirm that spacetime is dynamic, not static — capable of vibrating like the surface of a cosmic ocean.
They have opened a new window into the universe, allowing us to observe black hole collisions, neutron star mergers, and even the echoes of the Big Bang.

16.4 Relativity in Modern Technology and Cosmology

Though born from abstract reasoning, General Relativity underlies technologies we use every day — and shapes our understanding of the universe's origin and fate.

Global Positioning System (GPS)

GPS satellites orbit high above Earth, where gravity is weaker and time runs faster than on the ground.

However, they also move at high speeds, causing time dilation from Special Relativity.

Both effects — one speeding time, the other slowing it — must be precisely accounted for.
Without relativistic corrections, GPS positioning would drift by kilometers each day.
Thus, every smartphone and navigation device on Earth depends on Einstein's equations to function accurately.

Cosmology and the Expanding Universe

Einstein's field equations, when applied to the entire cosmos, form the foundation of **modern cosmology**. They predict that the universe cannot remain static — it must be **expanding or contracting**.

Observations by **Edwin Hubble** in 1929 confirmed that galaxies are receding from us, implying an **expanding universe** — just as Einstein's equations allowed.

Combined with quantum physics, General Relativity describes the **Big Bang**, the **evolution of cosmic structure**, and even the mysterious forces of **dark energy** and **dark matter** shaping the universe today.

The Curvature of the Cosmos

Depending on the total density of mass and energy, the universe's geometry can be:

- **Flat** (infinite and uncurved)

- **Closed** (finite and positively curved)

- **Open** (negatively curved and infinite)

Current observations suggest that the universe is remarkably flat, with a delicate balance between expansion and gravity — a testament to the precision of cosmic evolution.

16.5 The Legacy of General Relativity

Einstein's General Theory of Relativity is both a physical theory and a philosophical revelation. It redefined gravity not as a force, but as **geometry** — a manifestation of the curvature of space and time caused by mass and energy.

From the orbit of Mercury to the motion of galaxies, from GPS satellites to black holes and gravitational waves, every test confirms its truth.

Yet, at the frontier where relativity meets quantum mechanics — near singularities and in the early universe — new physics awaits. The quest for a unified **quantum theory of gravity** continues, inspired by Einstein's enduring vision.

16.6 The Harmony of Geometry and Reality

Einstein once wrote:

"The most incomprehensible thing about the universe is that it is comprehensible."

General Relativity reveals why: the universe is governed by an elegant geometry of motion, energy, and light. Spacetime bends, ripples, and expands — carrying within its structure the story of creation itself.

From falling apples to the warping of galaxies, from the ticking of atomic clocks to the merging of black holes, all are governed by the same geometric law.
It is a universe not of chaos, but of order — a cosmic symphony conducted by the curvature of spacetime.

Part VI – Quantum Mechanics: The Physics of the Microscopic World

Chapter 17: The Quantum Revolution

At the dawn of the 20th century, physics faced a crisis. The elegant laws of classical mechanics and electromagnetism could describe planets, projectiles, and even light waves — but they failed catastrophically when applied to the atomic world.

Why did hot objects glow in specific colors instead of continuously? Why did metals eject electrons only when illuminated by certain frequencies of light? Why were atomic spectra composed of sharp lines, not smooth bands?

These mysteries shattered the foundation of classical physics and gave rise to an entirely new way of understanding nature — **Quantum Mechanics**, the science of the very small.

It began with a series of bold ideas by **Max Planck, Albert Einstein**, and later **Louis de Broglie**, each challenging the assumption that energy and matter are continuous. Together, their discoveries formed the basis of the **Quantum Revolution** — a revolution that redefined light, matter, and the limits of human knowledge.

17.1 Planck, Einstein, and the Photon Concept

Planck's Quantum Hypothesis

In 1900, the German physicist **Max Planck** studied **blackbody radiation** — the spectrum of light emitted by a hot object. Classical theory predicted that as wavelength decreased, the emitted energy would increase without bound, leading to the absurd "ultraviolet catastrophe."

To resolve this, Planck made a daring assumption: **energy is not continuous, but quantized.**
He proposed that electromagnetic energy could only be emitted or absorbed in discrete packets called **quanta**, each proportional to the frequency of radiation.

$$E = hf$$

Where:

- E = energy of a quantum (joules)

- f = frequency (hertz)

- h = Planck's constant = 6.626×10^{-34} J · s

This simple equation introduced a fundamental constant of nature — **Planck's constant** — and marked the birth of quantum theory.

Planck himself viewed his idea as a mathematical trick, not a revolution. But a few years later, Einstein would reveal its deeper truth.

Einstein and the Photon

In 1905, Einstein extended Planck's concept to explain the **photoelectric effect** — the emission of electrons from a metal surface when struck by light.

Classical theory said that light waves should gradually transfer energy to electrons, regardless of frequency, depending only on intensity. But experiments showed otherwise:

- Light below a certain frequency, no matter how intense, **produced no electrons**.

- Above that threshold frequency, even weak light **ejected electrons instantly**.

Einstein proposed that light consists of **individual quanta of energy** — later called **photons** — each with energy $E = hf$.
A photon hitting an electron transfers its energy in an all-or-nothing collision. If the photon's energy exceeds the **work function** (the energy binding the electron to the metal), the electron escapes with kinetic energy.

$$K_{max} = hf - \phi$$

Where:

- K_{max} = maximum kinetic energy of ejected electron

- ϕ = work function of the metal

- f = frequency of incident light

This experiment confirmed that **light behaves as a particle** under certain conditions — a radical departure from the purely wave-based view of electromagnetic radiation.

Einstein's explanation earned him the **Nobel Prize in Physics in 1921** and cemented the photon as a fundamental concept in physics.

17.2 The Photoelectric Effect

The photoelectric effect demonstrates the **quantum nature of light** — that energy comes in discrete units rather than a continuous flow.

Key Observations:

1. **Threshold Frequency:**
   No electrons are emitted below a specific frequency, regardless of intensity.
   → Energy of photons depends on frequency, not brightness.

2. **Instantaneous Emission:**
   Electrons are ejected immediately when light of sufficient frequency strikes the surface.
   → Energy is transferred in single quantum interactions, not gradual accumulation.

3. **Kinetic Energy Depends on Frequency:**
   Increasing light frequency increases electron energy, while increasing intensity increases the number of electrons emitted.
   → Confirms the particle-like behavior of light.

Experimental Confirmation

Photoelectric experiments by **Robert Millikan** (1916) and others precisely verified Einstein's predictions and measured Planck's constant, providing direct evidence of light's quantized nature.

This discovery was revolutionary. It showed that:

- **Energy is discrete, not continuous.**

- **Light behaves both as a wave and as a stream of particles.**

It shattered the classical boundary between waves and matter, leading to a new question:
If light can act like a particle, **can particles act like waves?**

17.3 Wave–Particle Duality and the de Broglie Hypothesis

The Dual Nature of Light and Matter

For centuries, scientists debated whether light was composed of waves or particles.

- Newton believed in light particles ("corpuscles").

- Huygens and later Maxwell proved its wave nature.

- Einstein showed it had particle-like behavior as photons.

This duality — wave and particle at once — was paradoxical yet undeniable.

In 1924, French physicist **Louis de Broglie** extended this concept further, proposing that **matter** itself also exhibits wave-like properties.

The de Broglie Hypothesis

De Broglie reasoned that if light (once thought a wave) could behave as a particle, perhaps particles (like electrons) could behave as waves.

He proposed that any moving particle has a **wavelength** given by:

$$\lambda = \frac{h}{p}$$

Where:

- λ = wavelength (m)

- h = Planck's constant

- $p = mv$ = momentum of the particle

This means every particle — from electrons to baseballs — has a wavelength. But for macroscopic objects, the wavelength is so tiny that it's undetectable. For subatomic particles, however, it's significant and measurable.

Electron Diffraction and Proof of Matter Waves

In 1927, **Davisson and Germer** at Bell Labs confirmed de Broglie's hypothesis experimentally.
When electrons were fired at a crystal, they produced **interference patterns**, just like light waves.

This was irrefutable proof that **matter has wave-like behavior**, depending on its momentum and energy.

17.4 The Quantum Revolution Unfolds

The discoveries of Planck, Einstein, and de Broglie shattered the classical worldview.
No longer could energy, light, or matter be seen as purely continuous or discrete. Nature behaved as both — a **wave–particle duality** beyond classical logic but fully consistent with experiment.

This new framework opened the door to the **quantum mechanics** of **Schrödinger**, **Heisenberg**, and **Dirac**, who would build a complete theory describing the behavior of atoms, molecules, and subatomic particles.

The **Quantum Revolution** not only explained the structure of matter but also laid the foundation for modern technology — from lasers and semiconductors to computers and quantum communication.

17.5 The Birth of a New Reality

Quantum mechanics revealed that the microscopic world is not deterministic, but **probabilistic** — governed by likelihoods rather than certainties.
The act of observation itself influences what is measured, blurring the line between the observer and the observed.

What began as an attempt to explain glowing metals and light's behavior led to a profound shift in philosophy: Reality, at its core, is quantized — a dance of possibilities shaped by energy, probability, and the fundamental constants of nature.

The Quantum Revolution marked the beginning of a new age — one that would illuminate the atom, power the digital world, and unlock the mysteries of existence itself.

Chapter 18: Quantum Principles and the Atom

At the heart of all matter lies the atom — once thought indivisible, later revealed to be a miniature universe of particles bound by invisible forces.
Yet the behavior of electrons within atoms defied the logic of classical physics. They did not orbit the nucleus like planets around the Sun, nor could their energy vary continuously.

The atomic world, scientists discovered, is quantized — governed by discrete energy levels, probabilistic motion, and fundamental principles that challenge our deepest intuitions.

This chapter explores how **Bohr's model** first captured the quantized structure of the atom, how **Schrödinger's equation** described it as a wave system, and how **quantum principles** like superposition, uncertainty, and tunneling revealed a universe built not of certainty, but of possibility.

18.1 The Bohr Model and Quantized Energy Levels

In 1913, Danish physicist **Niels Bohr** proposed a model of the hydrogen atom that combined classical mechanics with the emerging ideas of quantum theory.
Bohr's insight resolved a great mystery: why atoms emit

or absorb light only at specific frequencies — producing sharp spectral lines rather than continuous colors.

Bohr's Postulates

1. **Electrons move in circular orbits** around the nucleus under electrostatic attraction.

2. **Only certain orbits are allowed**, where the electron's angular momentum is an integer multiple of Planck's constant divided by 2π:

$$L = n\frac{h}{2\pi}$$

where $n = 1,2,3, ...$(the principal quantum number).

3. **Energy is emitted or absorbed** only when an electron transitions between allowed orbits:

$$E_{\text{photon}} = hf = E_i - E_f$$

Each orbit corresponds to a **quantized energy level**, and transitions between levels produce discrete spectral lines — matching experimental observations of hydrogen's emission spectrum.

Bohr's Energy Levels

For hydrogen:

$$E_n = -\frac{13.6 \text{ eV}}{n^2}$$

where E_n is the energy of the electron in the nth orbit.

- $n = 1$: ground state (lowest energy)
- $n = 2,3, ...$: excited states

When an electron falls from a higher to a lower level, it emits a photon whose energy equals the difference between the levels — producing light of a precise wavelength.

This explained the **Balmer series** of hydrogen, one of the great triumphs of early quantum theory.

Limitations of Bohr's Model

While successful for hydrogen, the Bohr model failed for multi-electron atoms and could not explain why orbits should be quantized in the first place.
A deeper, wave-based theory was needed — one that would describe not just energy levels, but the very **nature of electrons**.

That theory would emerge a decade later through **Erwin Schrödinger**.

18.2 Schrödinger's Equation and Wave Functions

In 1926, Austrian physicist **Erwin Schrödinger** developed a mathematical framework for the quantum world.
He treated electrons not as point-like particles orbiting

the nucleus, but as **wave functions** — continuous distributions of probability that describe where the electron is likely to be found.

The Schrödinger Equation

The **time-dependent Schrödinger equation** is the fundamental law governing quantum systems:

$$i\hbar\frac{\partial \Psi}{\partial t} = \hat{H}\Psi$$

Where:

- Ψ(psi) = wave function of the system

- $\hbar = \frac{h}{2\pi}$ = reduced Planck's constant

- $\hat{H}$ = Hamiltonian operator (total energy: kinetic + potential)

For most atomic problems, the **time-independent Schrödinger equation** is used:

$$\hat{H}\Psi = E\Psi$$

The solutions Ψ represent **standing waves** in space, and their corresponding energies E are **quantized** — matching Bohr's energy levels, but derived from first principles.

The Wave Function (Ψ)

The wave function $\Psi(x, y, z, t)$ contains all the information about a quantum system, but it does not represent a physical wave in space.

Instead, the **square of its magnitude**, $|\Psi|^2$, gives the **probability density** — the likelihood of finding a particle in a particular region.

$$P(x, y, z) = |\Psi(x, y, z)|^2$$

This probabilistic interpretation, introduced by **Max Born**, marked a fundamental shift in physics:

The laws of nature no longer predict *what will happen*, but *the probabilities* of what *might* happen.

Quantum Numbers

Solutions to the Schrödinger equation for the hydrogen atom produce discrete energy levels and a set of **quantum numbers** that describe each electron's state:

1. **Principal quantum number (n):** energy level (1, 2, 3, ...)

2. **Angular momentum quantum number (l):** shape of orbital

3. **Magnetic quantum number (m):** orientation in space

4. **Spin quantum number (s):** intrinsic angular momentum of the electron

Together, these numbers define the structure of the atom — the basis of the periodic table and all of chemistry.

18.3 Quantum Superposition and the Uncertainty Principle

Superposition

In quantum mechanics, a particle can exist in multiple states simultaneously — a phenomenon known as **superposition**.
A system's wave function can be a combination (or superposition) of several possible states:

$$\Psi = c_1\Psi_1 + c_2\Psi_2 + \cdots$$

where $c_1, c_2, \dots$ are complex probability amplitudes.

Only when measured does the system "collapse" into one definite state.
This principle underlies quantum interference, atomic transitions, and modern quantum computing.

Superposition defies classical logic but has been confirmed experimentally countless times — from electrons passing through two slits to photons entangling across vast distances.

Heisenberg's Uncertainty Principle

In 1927, **Werner Heisenberg** discovered a profound limit to what can be known about a quantum system.
The **uncertainty principle** states that certain pairs of physical properties — like position and momentum — cannot both be known with arbitrary precision.

$$\Delta x \, \Delta p \geq \frac{\hbar}{2}$$

Where:

- Δx = uncertainty in position

- Δp = uncertainty in momentum

The more precisely we know one, the less precisely we can know the other.

This is not due to limitations of instruments, but a fundamental property of nature.
At the quantum level, particles do not have exact positions or velocities until observed — they exist as probability clouds, not points.

18.4 Quantum Tunneling

One of the most surprising predictions of quantum mechanics is **tunneling** — the ability of particles to pass through potential barriers that, classically, they do not have enough energy to overcome.

In classical physics, if a particle's energy E is less than the height of a potential barrier V, it cannot cross.

But in quantum mechanics, the wave function extends into the barrier with a small but finite probability of emerging on the other side.

$$T \propto e^{-2\kappa L}$$

where κ depends on the particle's energy and L is the barrier width.

This means that even the impossible is sometimes possible — at least in the probabilistic world of quantum mechanics.

Real-World Applications of Tunneling

- **Nuclear Fusion in Stars:**
  Protons in the Sun's core overcome repulsive forces via tunneling, allowing nuclear fusion to occur.

- **Quantum Tunneling Microscopes:**
  Scanning tunneling microscopes (STM) rely on electron tunneling to map surfaces at the atomic level.

- **Semiconductors and Electronics:**
  Quantum tunneling enables transistors, diodes, and flash memory — the foundation of modern computing.

Quantum tunneling is not a curiosity — it is one of the core processes powering both stars and modern technology.

18.5 The Quantum Atom: From Certainty to Probability

In classical physics, the atom was a miniature solar system; in quantum mechanics, it is a realm of waves and probabilities.
Electrons no longer follow precise orbits but occupy **orbitals** — regions where they are most likely to be found. Energy is quantized, motion is probabilistic, and observation itself shapes reality.

Through the quantum principles of Bohr, Schrödinger, Heisenberg, and others, we discovered that the universe is not built on fixed trajectories, but on the mathematics of uncertainty and harmony.
From this microscopic world arise all the colors of light, the structure of matter, and the chemistry of life.

18.6 The Beauty of the Quantum World

The quantum world defies intuition yet defines existence. Its laws are not chaos, but a deeper kind of order — one that unites light and matter, energy and probability, in a cosmic dance of wave and particle.

Every electron, atom, and molecule vibrates to the rhythm of Schrödinger's equation, and every photon echoes Planck's quantum.

Together they reveal a profound truth: the universe, at its foundation, is not solid but **statistical**, not mechanical but **mathematical**, and not predictable but **beautifully uncertain**.

Chapter 19: Quantum Applications

Quantum mechanics, once a mysterious theory of atoms and probabilities, has become the foundation of modern technology.

Its principles shape the devices that illuminate our cities, store our data, and connect our world. From the precise coherence of lasers to the power of quantum computers and the strange unity of entangled particles, the quantum revolution continues to redefine what is possible.

In this chapter, we explore how the laws of the microscopic world give rise to the most advanced technologies of the macroscopic one.

19.1 Lasers and Semiconductors

The Quantum Origin of the Laser

The word **LASER** stands for *Light Amplification by Stimulated Emission of Radiation*.
The laser's operation is a direct consequence of the **quantization of energy** in atoms.

In 1917, **Albert Einstein** introduced the concept of **stimulated emission**, showing that when an excited atom encounters a photon with energy matching its transition energy, it can be induced to emit a second photon that is *identical* — same frequency, direction, and phase.
This leads to a cascade of coherent photons — perfectly synchronized waves of light.

Principle of Laser Operation

A laser system consists of three essential components:

1. **Active medium** – atoms, ions, or molecules that can be excited (e.g., ruby, gas, or semiconductor).

2. **Energy source (pump)** – supplies energy to raise atoms to excited states.

3. **Optical cavity** – mirrors at both ends reflect light back and forth, amplifying it by stimulated emission.

When enough atoms are in excited states (a condition called **population inversion**), stimulated emission dominates, and coherent light escapes through a partially reflective mirror.

The result is a **beam of light** that is:

- **Monochromatic** – one color (frequency)

- **Coherent** – waves in phase

- **Directional** – highly focused

- **Intense** – concentrated energy

Applications of Lasers

- **Communication:** Optical fibers transmit data at the speed of light.

- **Medicine:** Laser surgery and precision cutting.

- **Industry:** Engraving, welding, and measurement tools.

- **Science:** Atomic clocks, spectroscopy, and fusion research.

- **Everyday life:** Laser printers, barcode scanners, and optical drives.

From atomic transitions to global connectivity, the laser embodies the power of quantum principles made practical.

Semiconductors: The Quantum Engine of Modern Electronics

Semiconductors lie at the heart of nearly all modern technology. Their behavior arises directly from **quantum energy bands** — allowed and forbidden regions for electron motion in solids.

In a semiconductor:

- The **valence band** contains bound electrons.

- The **conduction band** allows free movement of electrons.

- The **band gap** between them determines how easily electrons can move.

By controlling this gap through **doping** (adding impurities), engineers create materials that can conduct electricity selectively — forming the basis of **diodes, transistors**, and **integrated circuits**.

Quantum Tunneling in Semiconductors

Electrons in semiconductors can **tunnel** through potential barriers, enabling devices such as:

- **Tunnel diodes** (ultrafast switches)

- **Flash memory** (charge storage through thin barriers)

- **Quantum dots** (nanoscale structures where electrons are confined, producing discrete energy levels like artificial atoms)

Semiconductors and lasers together power the information age — an age built entirely upon the principles of quantum physics.

19.2 Quantum Computing and Cryptography

The Need for Quantum Computing

Classical computers, based on transistors, process information as **bits** — each bit being either **0 or 1**. Quantum computers, by contrast, use **quantum bits**, or **qubits**, which can exist in **superpositions** of 0 and 1 simultaneously.

This enables quantum computers to perform certain calculations exponentially faster than classical machines, by exploring many possible states at once.

Qubits and Superposition

A qubit's state is described by a **wave function**:

$$| \psi \rangle = \alpha \, | \, 0 \rangle + \beta \, | \, 1 \rangle$$

where α and β are complex probability amplitudes satisfying:

$$| \alpha |^2 + | \beta |^2 = 1$$

This superposition allows a system of n qubits to represent 2^n possible states simultaneously — a massive leap in computational parallelism.

Quantum Entanglement in Computation

Entanglement — the quantum correlation between particles — allows qubits to share information instantaneously across distance, acting as a unified system.
When entangled qubits are measured, their states are correlated no matter how far apart they are, enabling powerful computational and communication schemes.

Quantum Algorithms

Quantum algorithms exploit superposition and entanglement to achieve extraordinary speedups. For example:

- **Shor's algorithm**: factors large numbers exponentially faster than classical methods — threatening classical encryption systems.

- **Grover's algorithm**: searches unsorted databases in $\sqrt{N}$ time — faster than any classical approach.

Quantum computers could revolutionize fields such as **cryptography, optimization, machine learning**, and **materials science**.

Quantum Cryptography

While quantum computing may break classical encryption, it also enables **unbreakable communication** through **Quantum Key Distribution (QKD)**.

QKD relies on the fact that measuring a quantum system disturbs it — a direct consequence of the **uncertainty principle**.

In the **BB84 protocol**, for example, two parties exchange quantum bits (photons) to establish a shared key. Any attempt to eavesdrop introduces detectable errors, ensuring secure communication.

Quantum cryptography thus transforms uncertainty from a limitation into a shield of security.

19.3 Quantum Entanglement and Teleportation

The Mystery of Entanglement

In 1935, Einstein, Podolsky, and Rosen described a thought experiment — the **EPR paradox** — to highlight what they called "spooky action at a distance."
Two particles can become **entangled**, sharing a single quantum state even when separated by vast distances. Measuring one instantly determines the state of the other, regardless of space between them.

Modern experiments have confirmed entanglement as a real and fundamental aspect of nature — not an illusion or hidden connection, but an intrinsic nonlocal correlation predicted by quantum mechanics.

Entanglement violates no physical law, but it challenges classical ideas of **separability** and **local realism**, suggesting that the universe at its deepest level is **interconnected**.

Quantum Teleportation

Quantum teleportation, first demonstrated in 1997, uses entanglement to transmit the **quantum state** of a particle from one place to another — without moving the particle itself.

The process involves:

1. **An entangled pair** of particles shared between sender (Alice) and receiver (Bob).

2. **Measurement and classical communication** by Alice after interacting her particle with the state to be sent.

3. **Reconstruction** by Bob using Alice's data, reproducing the original quantum state perfectly on his side.

Teleportation does not transmit matter or energy instantaneously; it transfers *information* — the quantum blueprint — using a combination of quantum and classical channels.

This phenomenon is at the core of developing **quantum networks**, **secure communication systems**, and the future **quantum internet**.

19.4 The Quantum Frontier

Quantum technology stands where classical physics once did before the industrial age — on the threshold of transforming every field of human endeavor.

- **Quantum sensors** detect gravitational waves, magnetic fields, and time intervals with unprecedented precision.

- **Quantum materials** promise superconductors and energy-efficient devices.

- **Quantum communication** aims to create global networks immune to hacking.

- **Quantum computers** seek to solve problems beyond the reach of any classical machine.

The strange principles once confined to laboratories — superposition, entanglement, tunneling — now guide the technologies shaping the future.

19.5 The New Age of Quantum Reality

Quantum mechanics has evolved from a theoretical curiosity into the **engine of progress**.
From the glow of lasers to the logic of processors, from the mysteries of entanglement to the dreams of quantum computation, it teaches us that the universe's smallest scales are also its most powerful.

Where classical physics gave us the machine age, quantum physics has given us the **information age** — and perhaps, in time, the **quantum age**, where reality itself becomes programmable.

Part VII – Nuclear and Particle Physics: The Inner Workings of Matter

Chapter 20: The Atomic Nucleus

Deep within every atom lies a dense, compact core — the **nucleus**, the heart of matter.

Here, forces a hundred times stronger than those binding electrons hold together protons and neutrons in an extraordinarily small space.

The study of the nucleus reveals the origins of atomic energy, radioactivity, and the very elements that make up the universe. It bridges the microscopic and the cosmic — from the stability of matter to the fusion of stars.

20.1 Protons, Neutrons, and Isotopes

The Structure of the Nucleus

The **nucleus** resides at the center of the atom and contains nearly all its mass. It is composed of two types of particles known collectively as **nucleons**:

- **Protons (p^+):** Positively charged particles, each with a charge of $+1.602 \times 10^{-19}$ C.

- **Neutrons (n^0):** Electrically neutral particles, with nearly the same mass as protons.

Both are bound together by the **strong nuclear force**, an attractive interaction that overcomes the natural repulsion between positively charged protons at very short distances (about 10^{-15} meters).

Atomic Number and Mass Number

Each element is defined by its **atomic number (Z)** — the number of protons in its nucleus.
The **mass number (A)** is the total number of protons and neutrons.

$$A = Z + N$$

where N = number of neutrons.

Example:
Carbon-12 has 6 protons and 6 neutrons → $Z = 6$, $A = 12$.

The chemical behavior of an atom depends on Z, while its nuclear stability depends on both Z and N.

Isotopes

Atoms of the same element that differ in neutron number are called **isotopes**.
They share the same chemical properties but may have different masses and nuclear stabilities.

Examples:

- **Hydrogen-1 (protium):** 1 proton, 0 neutrons

- **Hydrogen-2 (deuterium):** 1 proton, 1 neutron

- **Hydrogen-3 (tritium):** 1 proton, 2 neutrons (radioactive)

Many elements have both stable and unstable isotopes, and the unstable ones undergo **radioactive decay** to achieve stability.

20.2 Binding Energy and Mass Defect

Mass Defect

When protons and neutrons combine to form a nucleus, the total mass of the nucleus is slightly **less** than the sum of its individual particles.
This missing mass, called the **mass defect**, corresponds to the **binding energy** that holds the nucleus together.

$$\Delta m = (Zm_p + Nm_n) - m_{nucleus}$$

The lost mass is converted into binding energy according to Einstein's equation:

$$E_b = \Delta mc^2$$

Binding Energy

Binding energy is the energy required to separate a nucleus into its individual nucleons.
It measures the strength of the nuclear force and is typically expressed in **mega-electronvolts (MeV)**.

A high binding energy per nucleon means a more stable nucleus.

Iron-56, for instance, has one of the highest binding energies per nucleon (~8.8 MeV), making it one of the most stable elements in the universe.

The Curve of Binding Energy

Plotting binding energy per nucleon against mass number A reveals a characteristic curve:

- Light nuclei (like hydrogen) can **fuse** to form heavier nuclei, releasing energy.

- Very heavy nuclei (like uranium) can **fission** into lighter nuclei, also releasing energy.

This curve explains why both **fusion** (in stars) and **fission** (in reactors) yield energy — both processes move nuclei toward greater stability.

20.3 Radioactive Decay: Alpha, Beta, and Gamma Radiation

Some isotopes are **unstable** because their nuclei contain an imbalance of protons and neutrons. These isotopes spontaneously transform into more stable forms by emitting radiation — a process known as **radioactive decay**.

Alpha (α) Decay

In **alpha decay**, the nucleus emits an **alpha particle** consisting of 2 protons and 2 neutrons (a helium-4 nucleus).

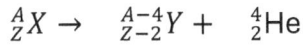

$$_{Z}^{A}X \rightarrow \; _{Z-2}^{A-4}Y + \; _{2}^{4}He$$

Alpha particles are relatively heavy and positively charged; they move slowly and have low penetrating power (stopped by paper or skin).

Example:

$$_{92}^{238}U \rightarrow \; _{90}^{234}Th + \; _{2}^{4}He$$

Beta (β) Decay

In **beta decay**, a neutron transforms into a proton (or vice versa), accompanied by the emission of a **beta particle** — an electron or positron — and a neutrino.

1. **Beta-minus (β⁻) decay:**

$$n \rightarrow p + e^- + \bar{v}_e$$

Example: Carbon-14 → Nitrogen-14.

2. **Beta-plus (β⁺) decay (positron emission):**

$$p \rightarrow n + e^+ + v_e$$

Occurs in proton-rich nuclei.

Beta particles are much lighter than alpha particles and have greater penetration, but can be stopped by a few millimeters of metal or plastic.

Gamma (γ) Radiation

After alpha or beta decay, a nucleus may remain in an **excited energy state**. It releases excess energy by emitting a **gamma ray** — a high-energy photon.

$$^{A}_{Z}X^* \rightarrow \ ^{A}_{Z}X + \gamma$$

Gamma radiation has no mass or charge and is extremely penetrating — requiring dense materials like lead or concrete for shielding.

Half-Life

Each radioactive isotope decays at a characteristic rate measured by its **half-life ($t_{1/2}$)** — the time it takes for half of a sample to decay.

$$N(t) = N_0 \left(\frac{1}{2}\right)^{t/t_{1/2}}$$

Half-lives range from fractions of a second (for unstable isotopes) to billions of years (for uranium-238).
This property allows scientists to date rocks, fossils, and even ancient artifacts using isotopic methods such as **carbon dating**.

20.4 The Power and Purpose of the Nucleus

The nucleus is both a source of danger and a wellspring of creation.
Its energy lights the stars through **fusion**, powers our technology through **fission**, and shapes the elements of life itself through **radioactive decay**.

Every element heavier than hydrogen was forged in stellar cores or supernovae — the universe's grand nuclear furnaces.

The same binding forces that stabilize matter also release the energy that drives the cosmos.

20.5 The Legacy of Nuclear Science

From the discovery of radioactivity by **Henri Becquerel**, to **Marie Curie's** isolation of radium, to the mastery of nuclear energy and medicine, the study of the atomic nucleus has transformed humanity's understanding of nature.

Yet it also carries profound responsibility — for the forces that hold the nucleus together are among the most powerful in existence.

In probing the nucleus, we glimpse both the **energy of creation** and the **potential for destruction** — a duality that mirrors our own capacity for wisdom and will.

Chapter 21: Nuclear Reactions

Within the heart of every atom lies a vast reservoir of energy — bound by the strong nuclear force that holds protons and neutrons together.

When this balance is disrupted, when nuclei split or merge, a portion of that binding energy is released in a single, awe-inspiring burst.

These processes — **nuclear fission** and **nuclear fusion** — are the engines of both stars and reactors.

They reveal the ultimate unity of matter and energy, first glimpsed in Einstein's famous equation:

$$E = mc^2$$

This chapter explores how humanity learned to harness these forces, how they power the cosmos, and how their energy sustains both life and technology.

21.1 Fission and Fusion

Nuclear Fission

Fission is the process in which a heavy nucleus splits into two (or more) lighter nuclei, releasing energy and additional neutrons.

It was first observed in 1938 by **Otto Hahn** and **Fritz Strassmann**, and explained by **Lise Meitner** and **Otto Frisch** using Einstein's mass–energy equivalence.

When a nucleus such as **uranium-235** absorbs a slow-moving neutron, it becomes unstable and splits into smaller nuclei — for example, **barium** and **krypton** — along with several free neutrons and vast amounts of energy:

$$^{235}_{92}U + {}^{1}_{0}n \rightarrow {}^{141}_{56}Ba + {}^{92}_{36}Kr + 3\ {}^{1}_{0}n + \text{Energy}$$

Each fission event releases around **200 MeV** of energy, primarily in the kinetic motion of the fragments.

Chain Reactions

The neutrons released in fission can strike other uranium nuclei, triggering more fissions — a **chain reaction**. If enough fissile material is present (the **critical mass**), the reaction becomes self-sustaining.

- **Controlled chain reaction:** Occurs in nuclear power plants, where control rods absorb excess neutrons.

- **Uncontrolled chain reaction:** Occurs in nuclear weapons, where the chain reaction grows exponentially.

Nuclear Fusion

Fusion is the opposite process — the joining of two light nuclei to form a heavier one.

In stars, fusion occurs under extreme temperatures and pressures, allowing atomic nuclei to overcome their electrostatic repulsion.

The most common reaction in the Sun is:

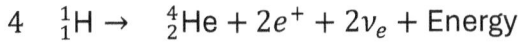

$$4 \ {}^1_1H \rightarrow \ {}^4_2He + 2e^+ + 2\nu_e + Energy$$

Here, hydrogen nuclei combine to form helium, releasing about **26.7 MeV** of energy per reaction.

Fusion releases even more energy per unit mass than fission, because it creates nuclei with higher binding energy per nucleon — moving closer to the peak of the binding energy curve.

Fission vs. Fusion

Property	Fission	Fusion
Typical nuclei	Uranium-235, Plutonium-239	Hydrogen, Deuterium, Tritium
Process	Splitting of heavy nucleus	Combining of light nuclei
Energy per reaction	~200 MeV	~400 MeV (per nucleon pair)
Byproducts	Radioactive fragments	Usually non-radioactive (helium)

Property	Fission	Fusion
Occurrence	Reactors, weapons	Stars, experimental reactors

Fusion promises **cleaner, safer, and more abundant** energy — the same power source that lights the stars.

21.2 Nuclear Energy and Power Plants

Controlled Fission in Reactors

Modern nuclear power plants harness fission through precisely managed chain reactions inside a **reactor core**.

Key components include:

- **Fuel rods:** Contain fissile material (usually U-235 or Pu-239).

- **Moderator:** Slows down neutrons to sustain fission (e.g., water, heavy water, or graphite).

- **Control rods:** Made of boron or cadmium to absorb excess neutrons and regulate reaction rate.

- **Coolant:** Transfers heat from the core to produce steam.

- **Turbine and generator:** Convert steam energy into electricity.

Nuclear Energy → Heat → Steam → Electricity

A single gram of uranium fuel can release as much energy as several tons of coal, with no direct carbon emissions — though it produces radioactive waste that must be safely stored.

Fusion Reactors: The Future of Energy

Fusion promises a nearly limitless, clean source of energy.
However, achieving and sustaining the required conditions — temperatures over **100 million K** — remains an immense scientific challenge.

The most promising approach uses **magnetic confinement** in devices called **tokamaks** or **stellarators**, which trap plasma using powerful magnetic fields.

Current international efforts, such as the **ITER project** in France, aim to demonstrate net-positive energy fusion within the coming decades.

Fusion reactions of deuterium and tritium are the focus:

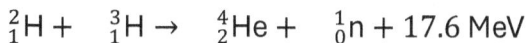

$$^{2}_{1}H + \ ^{3}_{1}H \rightarrow \ ^{4}_{2}He + \ ^{1}_{0}n + 17.6 \text{ MeV}$$

Fuel for such reactors — isotopes of hydrogen found in seawater — is abundant, making fusion the ultimate long-term energy goal for humanity.

21.3 Applications in Medicine and Astrophysics

Medical Applications

Nuclear physics has revolutionized modern medicine. Controlled radiation and isotopes are used to **diagnose**, **treat**, and **study** disease.

1. **Radiation Therapy:**
   High-energy gamma rays or particle beams target and destroy cancer cells while sparing surrounding tissue.

2. **Medical Imaging:**

   - **PET (Positron Emission Tomography)** uses positron-emitting isotopes (like fluorine-18) to visualize metabolic activity.

   - **MRI (Magnetic Resonance Imaging)** relies on nuclear spin interactions to map soft tissue in exquisite detail.

3. **Radioisotope Tracers:**
   Small amounts of radioactive substances track biological pathways in the body, aiding in diagnosis and research.

The same nuclear processes that power stars also illuminate the mysteries of life — showing how deeply connected the cosmos and biology truly are.

Astrophysical Applications

In the universe, nuclear reactions are the engines of cosmic evolution.

1. **Stellar Fusion:**
   Stars shine by fusing hydrogen into helium. As they age, heavier elements form in successive stages — carbon, oxygen, silicon, and iron.

2. **Supernovae and Element Formation:**
   When massive stars explode, extreme conditions fuse elements heavier than iron, scattering them across space.
   Every atom of gold, silver, and uranium on Earth was born in these titanic nuclear events.

3. **Cosmic Rays and Nucleosynthesis:**
   High-energy particles traveling through space continue to create isotopes in the interstellar medium, sustaining the cycle of cosmic chemistry.

Through fusion, the universe transforms matter into light — the same process that allows life to exist on our small blue planet.

21.4 The Power and Responsibility of Nuclear Science

Nuclear reactions represent both humanity's greatest scientific achievement and its greatest ethical challenge. They grant us the ability to **power cities** or **destroy them**, to **heal disease** or **harm the environment**.

The same principles that light the stars also remind us of the responsibility that comes with knowledge.
To understand the nucleus is to understand the creative

and destructive potential woven into the fabric of nature itself.

21.5 The Energy of the Universe

From the fusion furnaces of the Sun to the controlled reactors of Earth, from the birth of the elements to the radiance of distant galaxies, nuclear reactions are the pulse of the cosmos.
They unite energy and matter, past and future, destruction and creation — all within the smallest structures of existence.

In mastering the nucleus, humanity has learned not only how the universe generates its power, but also how that power reflects the profound balance between knowledge and wisdom.

Chapter 22: The Standard Model of Particle Physics

In humanity's quest to understand nature, the atom was once thought indivisible — the ultimate unit of matter. But deeper exploration revealed that atoms contain nuclei, nuclei contain protons and neutrons, and even these are composed of smaller constituents.

At the smallest scales accessible to experiment, all matter and energy are described by a remarkably elegant and successful framework known as the **Standard Model of Particle Physics**.

This theory unites the building blocks of matter — **quarks** and **leptons** — with the fundamental forces that govern their behavior — mediated by particles called **bosons**. Together, they describe the known structure of the universe with astonishing precision.

Yet, even this masterpiece is incomplete. The Standard Model explains almost everything we observe — except what lies **beyond**: the dark matter, dark energy, and deeper symmetries that still elude us.

22.1 Quarks, Leptons, and Bosons

The Families of Matter

All matter in the universe is composed of just **twelve fundamental particles** grouped into two main families:

quarks and **leptons**. These are the indivisible building blocks from which everything is made.

Family	Particle Type	Examples	Charge
Quarks	Up (u), Down (d), Charm (c), Strange (s), Top (t), Bottom (b)	Combine to form protons and neutrons	Fractional (+2/3 or – 1/3)
Leptons	Electron (e^-), Muon (μ^-), Tau (τ^-), and their neutrinos (v_e, v_μ, v_τ)	Fundamental particles, not made of quarks	Whole (–1 or 0)

- **Proton:** composed of 2 up quarks + 1 down quark → charge = +1

- **Neutron:** composed of 1 up quark + 2 down quarks → charge = 0

Each particle has an **antiparticle** with opposite charge and quantum properties (e.g., the **positron** is the electron's antiparticle).

The Force Carriers: Bosons

While quarks and leptons make up matter, their interactions are mediated by **force carriers** — particles known as **bosons**.

Force	Carrier (Boson)	Symbol	Acts On
Electromagnetic	Photon	γ	Charged particles
Weak nuclear	W^+, W^-, Z^0 bosons	$W^\pm, Z^0$	Quarks and leptons
Strong nuclear	Gluon	g	Quarks (binds inside nucleons)
Gravitational (hypothetical)	Graviton	G	All matter and energy

The **Higgs boson,** discovered in 2012, plays a unique role: it gives mass to other particles through the **Higgs field,** a universal quantum field that permeates all space.
Without it, particles would move at light speed and atoms could not exist.

22.2 Fundamental Interactions

Nature operates through **four fundamental interactions,** each with distinct strengths, ranges, and mediators.

Force	Relative Strength	Range	Carrier	Role
Strong	1 (strongest)	$\sim 10^{-15}$ m	Gluon	Binds quarks in protons, holds nuclei together
Electromagnetic	10^{-2}	Infinite	Photon	Governs light, electricity, magnetism
Weak	10^{-6}	$\sim 10^{-18}$ m	W^+, W^-, Z^0	Responsible for beta decay and neutrino interactions
Gravitational	10^{-38}	Infinite	Graviton (hypothetical)	Governs large-scale structure of the universe

Unification of Forces

Throughout history, physicists have discovered that seemingly distinct forces are different aspects of a deeper unity:

- Electricity and magnetism were unified by **Maxwell**.

- The weak and electromagnetic forces were unified in the **Electroweak Theory**, developed by **Glashow, Weinberg, and Salam**.

- The next great goal is the **Grand Unified Theory (GUT)**, which seeks to merge the strong, weak, and electromagnetic interactions — a step toward a single, all-encompassing **Theory of Everything (TOE)**.

22.3 Particle Accelerators and Detectors

To probe the subatomic world, scientists must recreate the high energies of the early universe — energies that allow matter to reveal its fundamental structure.

Particle Accelerators

Accelerators use electric and magnetic fields to speed particles to near-light velocities before smashing them together. The resulting collisions produce new particles, often lasting only fractions of a second.

Types include:

- **Linear accelerators (linacs):** Straight paths; used for precision experiments and medical therapy.

- **Circular accelerators (synchrotrons):** Particles move in loops, gaining energy with each pass.

  - Example: The **Large Hadron Collider (LHC)** at CERN — a 27 km ring beneath the French–Swiss border capable of 13 TeV collisions.

These collisions allow physicists to test theories, discover new particles, and study conditions similar to those after the Big Bang.

Particle Detectors

When high-energy collisions occur, the resulting particles leave **traces** that can be captured and analyzed using sophisticated detectors.

Major types include:

- **Cloud chambers** and **bubble chambers:** Visualize particle paths through ionized gas or liquid.

- **Scintillation detectors:** Convert particle energy into flashes of light.

- **Calorimeters:** Measure energy by absorption.

- **Tracking detectors** (like ATLAS and CMS at the LHC): Reconstruct 3D particle trajectories using electric and magnetic fields.

Through these technologies, the invisible architecture of the universe becomes visible — patterns of tracks, flashes, and energies that reveal the building blocks of reality.

22.4 Beyond the Standard Model

The Standard Model is a triumph — yet it is not the final word.
It successfully explains the behavior of known particles and forces but leaves profound mysteries unresolved.

Supersymmetry (SUSY)

Supersymmetry proposes that every known particle has a **superpartner** with a different spin:

- Each **fermion** (matter particle) has a **bosonic** partner.

- Each **boson** (force carrier) has a **fermionic** partner.

This elegant symmetry could unify forces at higher energies and explain why the Higgs boson is relatively light.
Although no superpartners have yet been observed, ongoing experiments continue to search for them.

String Theory

String theory suggests that all fundamental particles are not point-like but tiny, vibrating **strings** of energy. Different vibration modes correspond to different particles — like musical notes on a cosmic instrument.

It offers a framework that naturally includes **gravity**, making it a leading candidate for a **Theory of Everything**. However, its extra dimensions and vast mathematical complexity make direct testing extraordinarily difficult.

Dark Matter and Dark Energy

Observations of galaxies and the cosmic microwave background reveal that:

- Ordinary matter makes up only **5%** of the universe.

- **Dark matter** (~27%) interacts through gravity but not light — possibly composed of undiscovered particles like **WIMPs** or **axions**.

- **Dark energy** (~68%) drives the accelerating expansion of the universe, its nature still unknown.

The Standard Model describes only the **visible** universe — a mere fraction of cosmic reality.
Beyond it lies the next frontier of physics: uncovering the dark, invisible framework that shapes the cosmos.

22.5 The Quest for Unification

The story of physics has always been one of **unification**
— bringing together the many into the one.
From Newton's gravity to Maxwell's electromagnetism,
from Einstein's spacetime to the Standard Model's
symmetry, each step reveals a deeper simplicity behind
apparent complexity.

Physicists now seek the ultimate synthesis: a theory that
unites **quantum mechanics** with **general relativity**,
matter with energy, the very small with the very large.
Whether through **string theory**, **loop quantum gravity**, or
new ideas yet to come, the search continues.

22.6 The Architecture of the Universe

At the deepest level, reality is a web of energy and fields
— quarks and leptons interacting through fundamental
forces, woven together by symmetries and constants of
nature.
The Standard Model stands as one of humanity's greatest
intellectual achievements — a theory that describes
everything we can see, touch, and measure.

Yet, it may be only a fragment of a grander design — a
chapter in the universe's unfolding story, still being
written in the language of mathematics and imagination.

Part VIII – Astrophysics and Cosmology: The Universe and Beyond

Chapter 23: The Life and Death of Stars

From the faintest red dwarfs to the brightest supergiants, **stars** are the beating hearts of the cosmos.
They illuminate galaxies, forge the elements of life, and drive the evolution of the universe itself.
Yet every star, no matter how brilliant, has a beginning and an end — a story written in gravity, fusion, and time.

This chapter explores how stars form, how they shine through the power of nuclear fusion, and how they die — in quiet fades or cataclysmic explosions that seed the universe with the matter from which we are made.

23.1 Stellar Formation and Evolution

The Birth of Stars

Stars are born from vast clouds of **gas and dust** called **nebulae**, scattered throughout galaxies.
When regions of these clouds become dense enough, gravity pulls material inward, forming a **protostar** — a hot, collapsing sphere of hydrogen and helium.

As the protostar contracts, its temperature and pressure rise. When the core temperature reaches about **10 million kelvins**, hydrogen nuclei begin to fuse into helium.
At that moment, a star is born — entering the **main sequence** phase of its life.

Hydrostatic Equilibrium

A star's life is defined by balance.

- **Gravity** pulls inward, trying to collapse the star.

- **Pressure** from nuclear fusion pushes outward, counteracting gravity.

This delicate equilibrium maintains a star's stability for most of its lifetime.
If fusion increases, the star expands; if it decreases, the star contracts — a self-regulating cosmic harmony.

The Main Sequence

During the **main sequence**, a star steadily fuses hydrogen into helium in its core.
The star's mass determines its color, brightness, and lifespan:

- **Massive stars** burn hotter and brighter but live shorter lives (millions of years).

- **Smaller stars**, like red dwarfs, shine dimly for trillions of years.

- **Medium-mass stars**, like our Sun, live for about 10 billion years.

As hydrogen depletes, the star's internal balance begins to shift — marking the end of the main sequence and the beginning of stellar aging.

23.2 Nuclear Fusion in Stars

The Source of Stellar Energy

The incredible energy of stars comes from **nuclear fusion** — the joining of light nuclei into heavier ones.
In the Sun and similar stars, hydrogen nuclei (protons) fuse to form helium through the **proton–proton chain**:

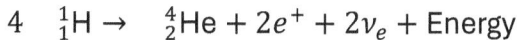

$$4 \ {}^{1}_{1}\text{H} \rightarrow \ {}^{4}_{2}\text{He} + 2e^{+} + 2\nu_{e} + \text{Energy}$$

Each reaction releases energy because the mass of the resulting helium nucleus is slightly less than the mass of the four protons.
That "missing" mass is converted into energy according to Einstein's equation, $E = mc^{2}$.

Fusion Cycles in Different Stars

1. **Proton–Proton Chain:**
   Dominant in smaller stars like the Sun. Converts hydrogen into helium.

2. **CNO Cycle (Carbon–Nitrogen–Oxygen):**
   In more massive, hotter stars, hydrogen fusion proceeds through a catalytic cycle involving carbon, nitrogen, and oxygen nuclei.

3. **Helium Fusion (Triple-Alpha Process):**
   When the hydrogen in the core is exhausted, the star contracts until helium can fuse:

$$3 \; {}^{4}_{2}\text{He} \rightarrow \; {}^{12}_{6}\text{C} + \text{Energy}$$

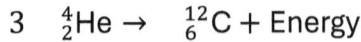

This process creates **carbon**, one of the essential elements for life.

4. **Advanced Fusion (in Massive Stars):**
   As the star evolves, it fuses heavier elements — oxygen, neon, silicon — building up layers like an onion, until iron forms at the core.
   Fusion of iron consumes energy rather than releasing it, signaling the star's imminent collapse.

23.3 The Death of Stars

Stars end their lives in profoundly different ways, depending on their mass.

Low- and Medium-Mass Stars

When a Sun-like star exhausts its hydrogen fuel, fusion ceases in the core. Gravity causes the core to contract while the outer layers expand, forming a **red giant**. Eventually, the star's outer layers drift into space, creating a glowing **planetary nebula**, while the core remains as a small, dense **white dwarf** — roughly the size of Earth but with the mass of the Sun.

White dwarfs no longer sustain fusion; they slowly cool and fade over billions of years, eventually becoming **black dwarfs** — cold, inert remnants of stellar light.

Massive Stars and Supernovae

Stars much larger than the Sun follow a far more violent fate.

Once iron accumulates in their cores, fusion can no longer produce energy to counter gravity. The core collapses in milliseconds, causing the outer layers to rebound in a colossal **supernova explosion**.

In this explosion:

- Elements heavier than iron — gold, uranium, and others — are forged in the intense heat and pressure.

- The shockwave disperses these elements into space, enriching future generations of stars and planets.

Supernovae are among the most energetic events in the universe — releasing more energy in a few seconds than the Sun will emit in its entire lifetime.

Neutron Stars

After a supernova, the collapsing core may compress so tightly that protons and electrons merge into **neutrons**. The result is a **neutron star** — an object only about 20 kilometers in diameter but containing more mass than the Sun.

A teaspoon of neutron star material would weigh billions of tons on Earth.

Many neutron stars rotate rapidly, emitting beams of radiation — observed as **pulsars**, cosmic lighthouses sweeping across space.

Black Holes

If the remnant core is more than about three times the Sun's mass, even neutron pressure cannot halt the collapse.

The core shrinks beyond its **Schwarzschild radius**, forming a **black hole** — a region of spacetime where gravity is so intense that nothing, not even light, can escape.

$$r_s = \frac{2GM}{c^2}$$

The surface boundary of no return is called the **event horizon**.

Inside, all known laws of physics break down, and matter collapses toward a singularity — a point of infinite density.

Black holes are detected indirectly, through the gravitational pull on nearby stars or the powerful **X-rays** emitted by infalling matter.

They are now known to reside not only at the ends of massive stars but also at the centers of galaxies, where

supermassive black holes shape the evolution of entire star systems.

23.4 The Cosmic Cycle of Life and Death

The death of one generation of stars gives birth to the next.
When a supernova explodes, its material enriches surrounding gas clouds with heavy elements — the raw ingredients of planets and life.

The carbon in our bodies, the oxygen we breathe, the calcium in our bones — all were forged in the fiery hearts of ancient stars and scattered across space by their deaths.

As the astronomer Carl Sagan once said, *"We are made of star stuff."*
The universe recycles itself endlessly, transforming matter into energy and energy into matter — an eternal cycle of creation and renewal.

23.5 The Legacy of Stellar Physics

Understanding stars unites every branch of physics — thermodynamics, quantum mechanics, nuclear physics, and relativity.
They are both laboratories and storytellers, revealing how the same laws that govern atoms also shape galaxies and cosmic history.

In studying the life and death of stars, we uncover our own origins — for the story of every atom in our bodies began in the heart of a star that lived, burned, and died long before the Sun was born.

Chapter 24: Galaxies and the Large-Scale Structure of the Universe

If stars are the atoms of the cosmos, then galaxies are its molecules — vast systems of stars, gas, dust, and dark matter bound together by gravity.
Each galaxy is a cosmic city containing hundreds of billions of stars, and there are hundreds of billions of galaxies scattered throughout the observable universe.

Together, they form a **cosmic web** — a network of filaments and voids spanning tens of billions of light-years, revealing the universe's hidden geometry.

This chapter explores how galaxies form and evolve, how dark matter shapes their motion, and how they gather into clusters and superclusters — the great architecture of creation itself.

24.1 Types of Galaxies and Their Dynamics

Classification of Galaxies

Galaxies come in many shapes and sizes, but they can be broadly classified into three main types — a system first developed by **Edwin Hubble** in the 1920s, known as the **Hubble Sequence** or **Tuning Fork Diagram**.

Type	Description	Example
Elliptical (E)	Smooth, featureless spheroids; little gas or dust; composed mostly of old stars.	M87 in Virgo Cluster
Spiral (S / SB)	Flat rotating disks with spiral arms; contain gas, dust, and young stars; central bulge of older stars.	Milky Way, Andromeda
Irregular (Irr)	Chaotic shapes with no defined structure; often the result of gravitational interactions or collisions.	Large Magellanic Cloud

- **Spiral galaxies** like our **Milky Way** are active sites of star formation, rotating in majestic spirals of glowing gas and dust.

- **Elliptical galaxies** are older, redder, and often found in dense clusters.

- **Irregular galaxies** are frequently distorted remnants of galactic mergers.

Structure and Motion of the Milky Way

Our galaxy, the **Milky Way**, is a barred spiral containing roughly **400 billion stars**.
It spans about **100,000 light-years** across and rotates once every **230 million years** — a cosmic year.

The Sun lies about 26,000 light-years from the galactic center, orbiting around a massive core that harbors a **supermassive black hole** known as **Sagittarius A***, with a mass of about **4 million Suns**.

The Milky Way is surrounded by a **halo** of old stars and **dark matter**, and is accompanied by smaller **satellite galaxies** such as the Magellanic Clouds.

Galaxy Formation and Evolution

Galaxies formed from **primordial density fluctuations** in the early universe — small irregularities in the cosmic microwave background that grew under gravity.
Over billions of years, these clumps attracted gas, cooled, and condensed into rotating disks where stars ignited.

Galaxies evolve through **star formation, mergers**, and **interactions**.
Collisions between galaxies are common — not as violent as they sound, since the distances between stars are immense — but gravitational forces can dramatically reshape galaxies, triggering new bursts of star formation.

Our Milky Way itself is on a collision course with the **Andromeda Galaxy (M31)**. In about **4.5 billion years**, they will merge into a new elliptical galaxy — sometimes called **Milkomeda**.

24.2 Dark Matter and Galactic Rotation Curves

The Mystery of Dark Matter

In the 1970s, astronomer **Vera Rubin** made a groundbreaking discovery while studying the rotation of spiral galaxies.
She found that stars far from the center of galaxies moved much faster than expected based on visible matter alone.

According to Newtonian gravity, rotational velocity should decrease with distance from the center — like planets in the solar system.
Instead, rotation curves **flattened**, suggesting that galaxies contain vast amounts of **invisible mass** extending far beyond their luminous disks.

This unseen mass became known as **dark matter** — a mysterious substance that does not emit, absorb, or reflect light, but reveals itself through its gravitational effects.

Evidence for Dark Matter

1. **Galaxy Rotation Curves:**
   Flat velocity profiles at large radii imply additional, unseen mass.

2. **Gravitational Lensing:**
   Light from distant galaxies bends more strongly than visible matter can explain.

3. **Cosmic Microwave Background (CMB):**
   Temperature fluctuations indicate a universe dominated by non-luminous matter.

4. **Galaxy Clusters:**
   The motion of galaxies within clusters requires far more mass than is visible.

Dark matter is thought to make up about **27% of the total mass–energy** of the universe — compared to only **5%** for ordinary matter.
The leading candidates are exotic, non-interacting particles such as **WIMPs (Weakly Interacting Massive Particles)** or **axions**.

The Dark Matter Halo

Every galaxy, including the Milky Way, is believed to be embedded in a vast, roughly spherical **dark matter halo**. This halo extends far beyond the visible edge of the galaxy, providing the gravitational glue that holds it together.

Without dark matter, galaxies would spin apart under their own rotation.
It is the unseen scaffolding of the cosmos — invisible, yet essential to its structure.

24.3 Clusters, Superclusters, and the Cosmic Web

Galaxy Clusters

Galaxies are not isolated; they gather into **clusters** bound by their mutual gravity.
Typical clusters contain hundreds or thousands of

galaxies, hot X-ray-emitting gas, and vast amounts of dark matter.

- **Virgo Cluster:** ~1,300 galaxies, including M87.
- **Coma Cluster:** One of the densest known, over 1,000 galaxies.

Within clusters, galaxies interact and merge, shaping their evolution.
Gravitational lensing by clusters distorts the light of distant galaxies, allowing astronomers to "see" the otherwise invisible dark matter through its gravitational imprint.

Superclusters and the Cosmic Web

Clusters themselves assemble into **superclusters** — the largest gravitationally bound structures in the universe. Our Milky Way belongs to the **Local Group**, which is part of the **Laniakea Supercluster**, containing over **100,000 galaxies** across 500 million light-years.

On even larger scales, galaxies and clusters form **filaments** and **walls**, separated by immense **voids** — regions nearly empty of galaxies.
These structures form a vast **cosmic web**, resembling neural networks or spider silk on the grandest imaginable scale.

Simulations and observations show that this web arose from tiny density fluctuations after the Big Bang, shaped over billions of years by the gravitational pull of dark matter.

Mapping the Universe

Modern telescopes and surveys — like the **Sloan Digital Sky Survey (SDSS)** and **James Webb Space Telescope (JWST)** — have mapped millions of galaxies, revealing the universe's large-scale structure in unprecedented detail.

The cosmic web stretches across the observable universe, spanning **tens of billions of light-years**. Yet, even this vastness may be but a fraction of the total — the visible surface of an ocean still largely unexplored.

24.4 The Architecture of the Cosmos

From stars to galaxies, from clusters to filaments, the universe exhibits hierarchy and harmony — a structure woven from both light and darkness. Gravity, guided by unseen matter, sculpts galaxies; nuclear fusion lights them from within; and their collective motion traces the hidden framework of spacetime.

The same laws that govern subatomic particles also govern the swirling of galaxies — physics uniting the infinitesimal with the infinite.

The universe, in its grandeur, is both a machine of matter and a masterpiece of mathematics — a living geometry stretching from quarks to clusters, from atoms to eternity.

Chapter 25: Cosmology and the Fate of the Universe

Cosmology is the science of the universe as a whole — its origin, structure, evolution, and ultimate fate.
It seeks to answer the most profound of all questions: *Where did everything come from? How did it evolve? And what will become of it all?*

Modern cosmology combines relativity, quantum mechanics, thermodynamics, and astrophysics into a single, breathtaking picture: a universe born in a moment of creation, expanding and evolving through time, governed by the same physical laws that shape atoms and galaxies alike.

25.1 The Big Bang Theory and Cosmic Background Radiation

The Birth of Space and Time

According to the **Big Bang theory**, the universe began approximately **13.8 billion years ago** from an extremely hot, dense state — a singularity where space, time, and energy were unified.

It was not an explosion in space, but an **expansion of space itself**. Every point in the universe moves away from every other point, as the fabric of spacetime stretches and carries matter along with it.

In the earliest moments (the **Planck Era**, 10^{-43} seconds after the beginning), all forces may have been unified. As the universe expanded and cooled, gravity, electromagnetism, and the nuclear forces separated, allowing particles, atoms, stars, and galaxies to form.

Evidence for the Big Bang

1. **Cosmic Expansion:**
   In 1929, **Edwin Hubble** observed that galaxies are moving away from us, and that their speeds increase with distance — expressed in **Hubble's Law:**

$$v = H_0 d$$

where v is recessional velocity, d is distance, and H_0 is the Hubble constant.
This discovery proved the universe is expanding.

2. **Cosmic Microwave Background (CMB):**
   Discovered accidentally by **Arno Penzias and Robert Wilson** in 1965, the CMB is the faint afterglow of the Big Bang — radiation left over from when the universe became transparent, about **380,000 years** after its birth.
   It fills all of space with a nearly uniform temperature of **2.73 K**, providing a direct window into the early cosmos.

3. **Abundance of Light Elements:**
   The predicted ratios of hydrogen, helium, and lithium formed during **Big Bang nucleosynthesis** match observations, confirming the theory's accuracy.

Together, these observations form overwhelming evidence that the universe began in a hot, dense state and has been expanding ever since.

25.2 Expansion of the Universe and Dark Energy

Hubble Expansion and the Cosmic Scale Factor

The universe's expansion is described by the **Friedmann–Lemaître–Robertson–Walker (FLRW)** metric, derived from Einstein's general relativity.
Galaxies are not moving through space; **space itself** is expanding, carrying galaxies apart.

This expansion is characterized by the **scale factor** $a(t)$— a function describing how distances between galaxies change with cosmic time. The **redshift** of light from distant galaxies measures this stretching of space.

Accelerating Expansion and Dark Energy

In 1998, two independent teams studying distant **Type Ia supernovae** made a shocking discovery: the universe's expansion is **accelerating**.

Something unseen is counteracting gravity — a mysterious form of energy dubbed **dark energy**.

Dark energy is estimated to make up about **68% of the total energy density** of the universe.
Its nature remains unknown, but it behaves like a **cosmological constant (Λ)** in Einstein's equations — a constant energy density filling all of space, causing spacetime itself to expand faster and faster.

$$H^2 = \frac{8\pi G}{3}\rho - \frac{kc^2}{a^2} + \frac{\Lambda c^2}{3}$$

Here, Λ represents the contribution of dark energy to the cosmic expansion.

The discovery of dark energy transformed cosmology — revealing that the ultimate fate of the universe depends not only on its matter content but on the mysterious energy of the vacuum itself.

25.3 The Ultimate Fate: Big Freeze, Big Crunch, or Big Rip

The universe's destiny depends on the interplay between **gravity** (which slows expansion) and **dark energy** (which accelerates it).
Several possible outcomes have been proposed:

1. The Big Freeze (Heat Death)

If the expansion continues forever, galaxies will drift farther apart until their light no longer reaches each other. Stars will exhaust their fuel, black holes will evaporate through **Hawking radiation**, and the universe will grow dark and cold.

In this scenario, energy becomes evenly distributed, and entropy — the measure of disorder — reaches its maximum.
This is the **heat death** of the universe: not an explosion, but a quiet fading into eternal darkness.

---

2. The Big Crunch

If the density of matter and energy is high enough, gravity could eventually halt the expansion and reverse it. Galaxies would begin to move toward one another, leading to a catastrophic collapse — a **Big Crunch** — where the universe ends as it began: in an infinitely dense singularity.

Some theories even propose that such a collapse could trigger a **new Big Bang**, suggesting the universe may cycle through endless expansions and contractions — a **cosmic rebirth**.

---

3. The Big Rip

If dark energy's repulsive force increases over time, it could tear apart the very fabric of the universe.

Galaxies, stars, planets, and even atoms would be pulled apart as the expansion accelerates without limit.

In this scenario, the universe ends not in cold or collapse, but in violent disintegration — the **Big Rip**.
Whether this will occur depends on the precise nature of dark energy, which remains one of physics' greatest mysteries.

25.4 The Multiverse Hypothesis

Beyond a Single Universe

Some theories suggest that our universe may not be alone.
The **multiverse hypothesis** proposes that our cosmos is one of many — perhaps infinitely many — universes, each with its own physical laws, constants, and dimensions.

There are several interpretations:

1. **Quantum Multiverse:**
   According to the **Many-Worlds Interpretation** of quantum mechanics, every quantum event creates new branches of reality — parallel universes in which every possible outcome occurs.

2. **Inflationary Multiverse:**
   During the early universe's rapid **inflation**, different regions may have stopped inflating at different times, creating "bubble universes" with different physical conditions.

3. **String Theory Landscape:**
   String theory predicts many possible vacuum states — potentially 10^{500} **universes**, each with its own laws of physics and dimensions.

4. **Brane Worlds:**
   In higher-dimensional theories, our universe may be a "brane" floating in a higher-dimensional space, interacting gravitationally with other branes — other universes.

While unproven, the multiverse concept challenges our deepest assumptions — suggesting that the cosmos we observe may be just one small island in a vast ocean of existence.

25.5 The Eternal Question

The study of cosmology brings us face-to-face with the ultimate mysteries — creation, time, existence, and destiny.
It reveals a universe that is both **comprehensible and incomprehensible**, governed by precise laws yet filled with profound wonder.

From the first spark of the Big Bang to the last fading photon, the cosmos evolves according to the same principles that govern the motion of atoms and galaxies alike.
To study the universe is to study ourselves — for we are its children, made of its matter, shaped by its history, and destined to share its fate.

Whether the universe ends in fire, ice, or transformation, one truth remains constant: the human spirit will always seek to understand.

For in exploring the universe, we discover not only its beginnings and endings, but the boundless curiosity that makes us part of its unfolding story.

Part IX – Modern Frontiers of Physics

Chapter 26: Unification and the Quest for a Theory of Everything

From the falling apple to the expansion of galaxies, physics seeks a common thread — a single set of principles that governs all things.
Throughout history, progress in physics has been a story of **unification**: bringing together phenomena once thought distinct under universal laws.

Newton unified the motion of the heavens and the Earth.
Maxwell unified electricity and magnetism.
Einstein unified space and time, and later mass and energy.

Today, the final challenge remains: to unify **quantum mechanics** and **general relativity**, the two great pillars of modern physics, into a single **Theory of Everything (TOE)** — a framework that explains the universe at every scale, from the smallest particle to the largest galaxy.

26.1 Grand Unified Theories (GUT)

The Dream of Unification

The **Standard Model** describes three of the four fundamental forces — electromagnetism, and the strong and weak nuclear forces — through the language of quantum field theory.
Gravity, however, stands apart, described by Einstein's

general relativity as the curvature of spacetime rather than a quantum field.

A **Grand Unified Theory (GUT)** aims to merge these three quantum forces into a single mathematical structure, reducing the number of fundamental interactions to one.

Gauge Symmetry and Unification

The Standard Model is built on **symmetries** — elegant mathematical transformations that leave physical laws unchanged.
GUTs propose that at extremely high energies (around 10^{16} GeV), these symmetries merge into a single larger symmetry group, often represented as:

$$SU(3) \times SU(2) \times U(1) \rightarrow SU(5) \text{ or } SO(10)$$

This unification implies that all forces were once the same in the early universe, only diverging as the cosmos cooled after the Big Bang.

GUTs predict that **protons may eventually decay**, albeit over timescales far longer than the current age of the universe — a prediction physicists continue to test with massive underground detectors.

Toward a Unified Force

If a GUT succeeds, it would combine the electromagnetic, weak, and strong forces into one interaction, described by

a single field.
The next step beyond this — incorporating **gravity** —
would yield the ultimate synthesis: the **Theory of
Everything,** where all forces, particles, and spacetime
itself emerge from one underlying principle.

26.2 String Theory and Quantum Gravity

The Problem of Quantum Gravity

Quantum mechanics describes the microscopic world
with exquisite precision, while general relativity governs
the cosmic scale.
Yet when we try to apply quantum theory to spacetime
itself — near singularities or inside black holes — the
equations break down.

This conflict between the **quantum** and the **gravitational**
is the greatest unresolved tension in physics.

Strings: The Music of the Universe

String theory offers a possible resolution.
It proposes that the fundamental constituents of nature
are not point-like particles, but tiny, vibrating **strings** of
energy.
Each vibration mode corresponds to a different particle —
like notes played on the same cosmic instrument.

In this view:

- A photon is one vibration mode.

- A quark is another.

- Even the graviton — the hypothetical quantum of gravity — arises naturally as a vibration of a closed string.

$$E = hf \text{ (energy of a vibrating string)}$$

Because strings have finite size, they smooth out the infinities that plague quantum gravity, allowing a consistent unification of all forces.

Extra Dimensions and M-Theory

For string theory to work mathematically, the universe must have more dimensions than the four we experience. Most versions require **10 spatial dimensions plus time**, while **M-theory**, a unifying extension of string theory, predicts **11 dimensions**.

The extra dimensions may be **compactified** — curled up so small they are imperceptible — or may exist as vast, hidden realms beyond our observable universe.

In this framework, our entire universe could be a **3-dimensional "brane"** floating in a higher-dimensional space, interacting gravitationally with other branes — perhaps other universes.

26.3 The Holographic Principle

Information and Reality

In the 1990s, physicist **Gerard 't Hooft** and later **Leonard Susskind** proposed the **holographic principle**, inspired by the physics of black holes.
According to this principle, **the information contained within a volume of space can be fully described by data encoded on its boundary surface** — much like a hologram encodes a 3D image on a 2D surface.

In essence, the universe may be **holographic**: the three-dimensional reality we experience could be a projection of fundamental information inscribed on a two-dimensional cosmic horizon.

Black Holes and Entropy

The idea arose from **Stephen Hawking's** work on black hole thermodynamics.
Black holes possess entropy proportional not to their volume, but to the **area of their event horizon**:

$$S = \frac{kA}{4l_P^2}$$

where A is the surface area and l_P is the Planck length.

This suggested that information — and therefore physical reality itself — may be stored not in space's interior, but on its boundaries.

The holographic principle thus bridges quantum theory, gravity, and information — hinting that **the universe itself may be a grand cosmic information system**.

26.4 Panpsychism and the Consciousness of the Universe

From Matter to Mind

While physics traditionally describes matter, energy, and forces, some thinkers propose that **consciousness** may be a fundamental property of the universe — not merely an emergent phenomenon of the brain.

This idea, known as **panpsychism**, suggests that **mind-like qualities** exist at every level of reality, from atoms to galaxies, forming an intrinsic aspect of the cosmic order.

The Universe as a Living System

In this interpretation, the universe is not a cold, mechanical construct, but a **living, evolving consciousness** — aware of itself through the minds it creates.
Each observer is a manifestation of the universe perceiving its own existence.

Modern physics indirectly echoes this idea:

- Quantum mechanics shows that **observation** influences outcomes.

- The holographic principle implies **information and perception** are central to reality.

- The laws of physics themselves seem fine-tuned to permit life and awareness.

Some scientists, such as **Roger Penrose** and **David Bohm**, have explored whether consciousness might be rooted in quantum processes — perhaps even woven into the fabric of spacetime.

A New Synthesis of Science and Mind

If the cosmos is both **physical and conscious**, then the final unification may not be only of forces, but of **existence itself** — matter, energy, information, and awareness as facets of a single universal reality.

The **Theory of Everything** might then be not just a set of equations, but a profound realization: that the universe is a self-aware system — **a cosmic mind reflecting upon itself**.

26.5 Toward the Ultimate Understanding

The pursuit of unification is not merely a scientific endeavor; it is the human spirit seeking its own reflection in the universe.
Whether through strings, fields, or consciousness, every step toward a deeper theory reveals the same truth —

that **the universe is one**, bound by symmetry, logic, and beauty.

The dream of a **Theory of Everything** is more than a quest for equations.
It is a quest to understand why the universe exists, why it is comprehensible, and why within its vastness, something like us can ask these questions at all.

In the unity of physics, mathematics, and consciousness, we may finally glimpse the ultimate harmony — the moment when **the observer and the observed become one**, and the universe awakens to its own infinite self.

Chapter 27: Physics in Technology and Society

Physics is not confined to laboratories or equations — it lives within every aspect of modern civilization.
From the power that lights our cities to the satellites orbiting above, from medical imaging to artificial intelligence, the principles of physics have transformed human existence.

Yet with this power comes responsibility — to use knowledge not only to advance technology, but to sustain the planet, explore the cosmos, and guide humanity toward a wiser future.

This chapter explores the profound connection between physics, technology, and society — and how the discoveries of today will shape the destiny of tomorrow.

27.1 Energy and Sustainability

The Physics of Energy

At its core, energy is the capacity to do work — and all forms of energy, from mechanical to nuclear, are governed by the same fundamental principles.
The **law of conservation of energy** ensures that energy can neither be created nor destroyed, only transformed.

Modern civilization depends on these transformations — converting chemical energy into electricity, sunlight into motion, and atomic energy into light and heat.

Renewable Energy and the Future

As humanity faces climate change and resource depletion, the challenge is to harness **sustainable** sources of energy that align with physical and environmental limits.

Key developments include:

- **Solar Power:** Converts sunlight into electricity via the **photoelectric effect**, first explained by Einstein in 1905.

- **Wind and Hydroelectric Energy:** Transform mechanical motion into electrical energy using electromagnetic induction.

- **Geothermal and Tidal Power:** Utilize Earth's internal heat and gravitational effects of the Moon and Sun.

- **Nuclear Fusion:** Offers the promise of near-limitless clean energy, mimicking the process that powers the stars.

Physics provides the foundation for all these innovations — from materials science for efficient solar panels to plasma physics for fusion reactors.

Entropy and Efficiency

According to the **Second Law of Thermodynamics**, no energy transformation is perfectly efficient — some energy is always dissipated as heat.
Understanding this law helps engineers design systems that maximize output while minimizing waste, essential for sustainable technology.

The future of energy lies not only in new sources, but in mastering **efficiency**, **storage**, and **balance** — harmonizing the laws of physics with the needs of life on Earth.

27.2 Space Exploration and Propulsion Physics

The Physics of Spaceflight

Space exploration is one of humanity's greatest triumphs — a direct application of Newton's laws, gravitation, and conservation of momentum.

Rockets operate by expelling mass in one direction to propel themselves in the opposite direction, as described by Newton's Third Law:

$$F = \frac{dmv}{dt}$$

Even in the vacuum of space, momentum exchange provides thrust.

Chemical and Electric Propulsion

1. **Chemical Rockets:**
   Use high-temperature combustion to eject gases at tremendous speeds. These remain the backbone of space travel, from the Saturn V to modern Falcon rockets.

2. **Ion and Plasma Propulsion:**
   Accelerate ions using electromagnetic fields for long-duration missions. Although thrust is low, efficiency and endurance are high — ideal for interplanetary exploration.

3. **Nuclear and Fusion Propulsion:**
   Concepts such as **nuclear thermal rockets** and **fusion drives** could enable travel to Mars and beyond, harnessing the immense energy density of atomic reactions.

4. **Photon and Antimatter Propulsion:**
   Theoretical technologies like **light sails** or **matter-antimatter engines** could one day reach relativistic speeds — turning science fiction into science fact.

The Future of Human Expansion

Einstein's relativity sets limits on speed and energy, but not on imagination.
Advances in propulsion physics may one day allow humanity to explore exoplanets or even build interstellar vessels powered by **fusion**, **antimatter**, or **warp-field concepts** derived from spacetime engineering.

The same physics that explains falling apples now guides ships among the stars.

27.3 Artificial Intelligence and Quantum Computing

The Physics of Intelligence

Artificial intelligence (AI) may seem like a branch of computer science, but its roots lie in **physics** — particularly in thermodynamics, information theory, and quantum mechanics.

Every bit of information processed by a computer carries physical consequences, consuming energy and generating entropy.
The relationship between **information and energy** — formalized by Rolf Landauer's principle — reveals that even thinking machines obey the laws of physics.

Quantum Computing

Quantum computing harnesses the strange principles of superposition and entanglement to process information in ways classical computers cannot.

A **qubit** can represent both 0 and 1 simultaneously:

$$| \psi \rangle = \alpha \, | \, 0 \rangle + \beta \, | \, 1 \rangle$$

This allows quantum computers to explore many possible outcomes in parallel.

Quantum gates manipulate these states through unitary transformations, performing computations exponentially faster for certain problems — like factorization, simulation, and optimization.

Applications of quantum computing include:

- **Cryptography** (quantum-safe security)

- **Material science and chemistry**

- **Artificial intelligence and pattern recognition**

- **Climate modeling and energy optimization**

Quantum computing represents the merging of information and physics — computation performed not just by machines, but by nature itself.

AI and Physics in Harmony

AI is revolutionizing physics, just as physics enables AI. Machine learning algorithms now analyze cosmic data, simulate quantum systems, and even aid in discovering new physical laws.
In turn, physics provides the principles — of computation, energy, and logic — that make artificial intelligence possible.

Together, they mark a new age: **the intelligent universe**, where the tools of physics become extensions of the human mind.

27.4 The Future of Physics and Human Civilization

The Technological Singularity and Beyond

As technology accelerates, some scientists foresee a **singularity** — a point where artificial intelligence surpasses human intellect, leading to rapid, unpredictable change.
Whether this marks transcendence or danger depends on humanity's wisdom in guiding its creations.

Physics, as the foundation of all technology, will remain our compass.
Understanding the limits of energy, entropy, and computation will help ensure that innovation aligns with the sustainability of life and the ethics of progress.

Interplanetary and Interstellar Civilizations

Physics defines the boundaries of possibility — and those boundaries are vast.
Through advances in propulsion, energy, and communication, humanity may one day become a **Type I** (planetary), **Type II** (stellar), or even **Type III** (galactic) civilization on the **Kardashev scale** — harnessing energy from planets, stars, and galaxies.

Each step will require mastery of physical law — from quantum control to cosmic engineering — and a deep respect for the balance between power and purpose.

The Moral Dimension of Science

With knowledge comes responsibility. The same physics that powers cities can power weapons; the same equations that guide spacecraft can design surveillance systems.
It is not physics itself that determines humanity's fate, but how we choose to apply it.

A sustainable, peaceful, and enlightened civilization will depend not only on understanding the universe, but also on understanding ourselves.

Toward the Next Horizon

Physics will continue to evolve — integrating with biology, information science, and philosophy — as humanity seeks ever deeper truths about the nature of existence. Perhaps the ultimate discovery will not be about matter or energy at all, but about **meaning** — the realization that the universe's greatest creation is consciousness aware of itself.

The journey of physics mirrors the journey of humanity: from curiosity to comprehension, from survival to wisdom, from observing the stars to reaching them.

27.5 The Legacy of Physics

Every law of motion, every quantum equation, every spark of electricity represents a triumph of the human spirit —

our relentless desire to understand.
Physics has taken us from the cave to the cosmos, from candlelight to fusion, from the mystery of the atom to the mystery of existence itself.

The universe continues to unfold before us, inviting exploration, imagination, and humility.
And as long as there are minds that ask *why*, the story of physics — and of humanity — will never end.

Part X – Appendices

Reference Materials and Supporting Knowledge

Appendix A: Fundamental Constants and Units

The constants and units below form the backbone of all physical calculations.
They represent nature's unchanging values — the "grammar" of the universe through which all physical laws are written.

Quantity	Symbol	Value	Unit (SI)
Speed of light in vacuum	c	2.99792458×10^8	m/s
Gravitational constant	G	6.67430×10^{-11}	$m^3/kg{\cdot}s^2$
Planck's constant	h	$6.62607015 \times 10^{-34}$	J·s
Reduced Planck constant	$\hbar = \dfrac{h}{2\pi}$	$1.0545718 \times 10^{-34}$	J·s
Elementary charge	e	$1.60217663 \times 10^{-19}$	C
Electron mass	m_e	$9.1093837 \times 10^{-31}$	kg

Quantity	Symbol	Value	Unit (SI)
Proton mass	m_p	$1.6726219 \times 10^{-27}$	kg
Neutron mass	m_n	$1.6749275 \times 10^{-27}$	kg
Avogadro's number	N_A	$6.02214076 \times 10^{23}$	mol^{-1}
Boltzmann constant	k_B	1.380649×10^{-23}	J/K
Gas constant	R	8.3144626	J/mol·K
Stefan–Boltzmann constant	σ	5.670374×10^{-8}	$W/m^2{\cdot}K^4$
Coulomb constant	$k_e = \dfrac{1}{4\pi\varepsilon_0}$	8.98755×10^9	$N{\cdot}m^2/C^2$
Permittivity of free space	ε_0	$8.8541878 \times 10^{-12}$	$C^2/N{\cdot}m^2$
Permeability of free space	μ_0	$4\pi \times 10^{-7}$	N/A^2
Standard acceleration due to gravity	g	9.80665	m/s^2
Astronomical unit	AU	1.4959787×10^{11}	m

Quantity	Symbol	Value	Unit (SI)
One light-year	—	9.4607×10^{15}	m
Planck length	l_P	1.616255×10^{-35}	m

Appendix B: Common Equations and Derivations

Mechanics

$F = ma$(Newton's Second Law)

$W = Fd\cos\theta$(Work)

$KE = \frac{1}{2}mv^2, PE = mgh$(Energy)

$p = mv, F = \frac{dp}{dt}$(Momentum and Impulse)

$\tau = rF\sin\theta, I = \sum mr^2$(Torque and Moment of Inertia)

$U_g = -\frac{GMm}{r}$(Gravitational Potential Energy)

Thermodynamics

$Q = mc\Delta T$(Heat Transfer)

$\Delta U = Q - W$(First Law of Thermodynamics)

$\eta = 1 - \frac{T_C}{T_H}$(Carnot Efficiency)

Electromagnetism

$$F = k_e \frac{q_1 q_2}{r^2} \text{(Coulomb's Law)}$$

$$V = \frac{U}{q}, E = -\nabla V \text{(Potential and Field)}$$

$$V = IR \text{(Ohm's Law)}$$

$$\Phi_B = BA, \mathcal{E} = -\frac{d\Phi_B}{dt} \text{(Faraday's Law)}$$

Waves and Optics

$$v = f\lambda \text{(Wave Speed)}$$

$$n_1 \sin \theta_1 = n_2 \sin \theta_2 \text{(Snell's Law)}$$

$$I = I_0 \cos^2 \theta \text{(Malus' Law)}$$

Relativity and Quantum Physics

$$E = mc^2, E^2 = (pc)^2 + (m_0 c^2)^2$$

$$\lambda = \frac{h}{p} \text{(de Broglie Wavelength)}$$

$$E = hf \text{(Photon Energy)}$$

$$\Delta x \Delta p \geq \frac{\hbar}{2} \text{(Heisenberg Uncertainty Principle)}$$

Appendix C: Greek Alphabet and Mathematical Symbols

Letter	Name	Common Usage in Physics
α	alpha	Angular acceleration, fine-structure constant
β	beta	Velocity ratio (v/c), beta particles
γ	gamma	Photon, Lorentz factor, gamma rays
δ	delta	Change or variation
Δ	capital delta	Finite change (Δx, ΔE)
ε	epsilon	Permittivity, small quantity
θ	theta	Angle
λ	lambda	Wavelength
μ	mu	Coefficient of friction, micro (10^{-6})
π	pi	Ratio of circumference to diameter (≈ 3.14159)
ρ	rho	Density, resistivity
σ	sigma	Summation, surface charge density, stress
τ	tau	Torque, time constant
φ	phi	Electric potential, magnetic flux
ω	omega	Angular frequency

Letter	Name	Common Usage in Physics
Ω	capital omega	Ohm (unit of resistance)
ψ	psi	Wave function in quantum mechanics

Appendix D: Historical Timeline of Major Discoveries

Year	Discovery / Event	Scientist(s)
1543	*Heliocentric model published*	Copernicus
1609–1619	Laws of planetary motion	Kepler
1687	*Principia Mathematica* – laws of motion, gravity	Newton
1803	Atomic theory	Dalton
1820	Electromagnetism discovered	Ørsted
1865	Maxwell's equations	James Clerk Maxwell
1896	Discovery of radioactivity	Becquerel, Curie
1900	Quantum hypothesis	Planck
1905	Special relativity; photoelectric effect	Einstein

Year	Discovery / Event	Scientist(s)
1911	Nuclear atom model	Rutherford
1915	General relativity	Einstein
1926–1927	Quantum mechanics formulated	Schrödinger, Heisenberg
1932	Discovery of the neutron	Chadwick
1942	First nuclear reactor	Fermi
1953	DNA structure (physics in biology)	Watson & Crick
1965	Cosmic microwave background discovered	Penzias & Wilson
1970s	Standard Model developed	Glashow, Weinberg, Salam
1989	Cold fusion claim (debunked, but influential)	Fleischmann & Pons
2012	Higgs boson discovered	CERN / LHC
2015	Gravitational waves detected	LIGO
2020s	Quantum computing and dark energy studies expand	Global collaboration

Appendix E: Glossary of Key Terms

Acceleration (a): The rate of change of velocity over time.

Amplitude: Maximum displacement in a wave.

Atom: The smallest unit of an element, composed of protons, neutrons, and electrons.

Black Hole: A region where gravity is so strong that not even light can escape.

Conservation Law: A principle stating that certain quantities (energy, momentum, charge) remain constant in isolated systems.

Dark Energy: A mysterious force causing the accelerated expansion of the universe.

Entropy: A measure of disorder or randomness in a system.

Force (F): A push or pull that can cause an object to accelerate.

Gravity: The attractive force between masses.

Kinetic Energy: Energy of motion.

Mass: A measure of the amount of matter in an object.

Photon: A quantum of light; the carrier of electromagnetic energy.

Quantum: The smallest discrete amount of a physical quantity.

Relativity: The theory describing the relationship between space, time, and motion.

Wave–Particle Duality: The concept that matter and energy can exhibit both wave-like and particle-like behavior.

(The full glossary would expand alphabetically with clear, concise definitions of all major physics terms.)

Appendix F: Further Reading and Resources

Foundational Texts

- *The Feynman Lectures on Physics* — Richard P. Feynman, Robert B. Leighton, and Matthew Sands

- *Six Easy Pieces* — Richard P. Feynman

- *The Elegant Universe* — Brian Greene

- *A Brief History of Time* — Stephen Hawking

- *The Character of Physical Law* — Richard P. Feynman

- *The Road to Reality* — Roger Penrose

Modern Physics and Cosmology

- *The Fabric of the Cosmos* — Brian Greene

- *Black Holes and Time Warps* — Kip Thorne

- *Cosmos* — Carl Sagan

- *The Universe in a Nutshell* — Stephen Hawking

Online Learning Resources

- **NASA**: https://www.nasa.gov

- **CERN**: https://home.cern

- **MIT OpenCourseWare – Physics**: https://ocw.mit.edu

- **Khan Academy – Physics:**
  https://www.khanacademy.org/science/physics

- **WolframAlpha Physics Engine:**
  https://www.wolframalpha.com

A Final Note

These appendices serve as both **a foundation and a springboard** — a place for quick reference and deeper exploration.

They connect the timeless constants of physics with the evolving curiosity of its students, reflecting the unity between precision and wonder that defines all scientific discovery.

Epilogue: The Endless Horizon of Knowledge

By Matthew Meyer

Physics began with a question — *Why?*

Why do the stars move as they do? Why does the apple fall? Why does light shine, and time flow, and existence unfold in such intricate harmony?
From that single word arose a discipline that would change not only our understanding of nature, but our understanding of ourselves.

Through the centuries, physics has been the story of human curiosity made manifest — of minds daring to look beyond the surface of things, to see the order behind the chaos and the unity behind the diversity.
Each discovery has peeled back another layer of reality, revealing that the universe is not a collection of disconnected parts, but a single, coherent whole — a cosmos woven from energy, matter, space, time, and perhaps, consciousness itself.

The Journey from Earth to Infinity

From the motions of falling bodies to the expansion of galaxies, physics has shown that the same laws apply everywhere — that the Earth and the heavens obey one language.
This realization transformed mystery into meaning: the

gravitational pull that guides an apple's fall is the same that binds the stars in their spirals.

We have learned to see the invisible — to uncover the quantum world beneath perception, where particles blur into probabilities and the boundaries between matter and energy dissolve.

We have journeyed through the birth of stars, the death of galaxies, and the whispers of the Big Bang — tracing the story of the universe from its fiery beginning to its possible ends.

And through it all, we have discovered that to know the universe is, in some sense, to know ourselves — for we are made of its dust, born from its light, and sustained by its laws.

From Matter to Mind

The deeper physics probes, the more it encounters the question of awareness.

How does a universe of particles and waves give rise to minds that can ponder them?

How does matter, governed by impersonal laws, awaken to self-awareness and ask the very questions that led to its own discovery?

Perhaps consciousness is not an accident, but a reflection — the universe seeing itself through our eyes. Perhaps every equation and experiment is part of a cosmic dialogue between matter and meaning, between the measurable and the mysterious.

Modern physics has reached the threshold where science and philosophy converge — where the quest for a *Theory of Everything* becomes not only a search for the unity of forces, but for the unity of existence.

We are participants in that search, not spectators. Every thought, every act of wonder, is part of the grand equation.

The Infinite Horizon

There is no final frontier of knowledge, only ever-widening horizons.

Each answer gives rise to new questions, and each mystery unfolds into deeper beauty.

This is not a failure of science, but its glory — for the universe is infinite in depth as well as in scale, and the joy of discovery is eternal.

The journey of physics mirrors the evolution of consciousness itself:

from curiosity to comprehension, from observation to insight, from the physical to the transcendent.

It reminds us that understanding the universe is not about reducing it to numbers, but about awakening to its majesty — a symphony of patterns, energies, and possibilities.

The Human Destiny

The story of physics is ultimately the story of humanity. With every equation written, every particle discovered, and every star observed, we affirm the power of reason, imagination, and hope.
We are not separate from the cosmos — we are its expression, its voice, its awareness.

The atoms in our bodies were forged in the hearts of ancient stars; the thoughts in our minds are shaped by the laws of nature.
To explore the universe is to return home — to rediscover that we belong to something vast, timeless, and beautiful.

The future of physics will not merely be written in laboratories or equations, but in the hearts of those who dare to dream — who see in the universe not just complexity, but meaning; not just motion, but purpose.

The Endless Horizon of Knowledge

And so, the journey continues.

Physics will evolve, our understanding will deepen, and the universe will reveal new wonders — perhaps never-ending ones.
Each generation will look up at the stars and ask again: *Why?*
And each answer, like a ripple in eternity, will bring us closer to the realization that the universe is not only out there — it is also within us.

The cosmos is vast, yet knowable.

Mysterious, yet harmonious.

And through our pursuit of knowledge, we become what we study — luminous minds reflecting a luminous universe.

For the truest equation of all may be this:

Consciousness = The Universe Awake to Itself.

The horizon of knowledge has no end — only beginnings. And as long as we seek to understand, the light of discovery will never fade.

Matthew Meyer

Author of Foundations of Physics: A Comprehensive Journey Through the Laws of the Universe

www.ingramcontent.com/pod-product-compliance
Lightning Source LLC
Chambersburg PA
CBHW070349200326
41518CB00012B/2185